FOUNDATION
REPAIR
MANUAL

FOUNDATION REPAIR MANUAL

Robert Wade Brown

*Brown Foundation Repair
and Consulting Services, Inc.*

McGraw-Hill

New York   San Francisco   Washington, D.C.   Auckland   Bogotá
Caracas   Lisbon   London   Madrid   Mexico City   Milan
Montreal   New Delhi   San Juan   Singapore
Sydney   Tokyo   Toronto

Library of Congress Cataloging-in-Publication Data

Brown, Robert Wade.
    Foundation repair manual / Robert Wade Brown.
      p. cm.
    ISBN 0-07-008244-8
    1. Foundations—Maintenance and repair—Handbooks, manuals, etc.
  I. Title.
  TA775.B672   1999
  624.1'5'0288—dc21   98-32430
                                              CIP

McGraw-Hill

A Division of The McGraw·Hill Companies

1 2 3 4 5 6 7 8 9 0 DOC/DOC 9 0 4 3 2 1 0 9

ISBN 0-07-008244-8

The sponsoring editor of this book was Larry S. Hager, the editing supervisor was Paul R. Sobel, and the production supervisor was Sherri Souffrance. It was set in the HB1 design in Times Roman by Michele M. Zito of McGraw-Hill's Professional Book Group composition unit.

Printed and bound by R. R. Donnelley & Sons Company.

McGraw-Hill books are available at special quantity discounts to use as premiums and sales promotions, or for use in corporate training programs. For more information, please write to the Director of Special Sales, McGraw-Hill, 11 West 19th Street, New York, NY 10011. Or contact your local bookstore.

This book was printed on recycled, acid-free paper containing a minimum of 50% recycled, de-inked fiber.

CONTENTS

Chapter 3. Mudjacking (Slab)   3.1

Chapter 4. Deep Grouting   4.1

Chapter 5. Underpinning   5.1

Chapter 6. Basement and Foundation Wall Repair   **6.1**

Chapter 7. Soil Stabilization   **7.1**

Chapter 11. Foundation Inspection and Property Evaluation for the Residential Buyer 11.1

Chapter 12. Case Histories 12.1

PREFACE

This book is the result of over 35 years activity and research devoted to analyzing and correcting failures in foundations. Although foundation design per se is considered to be beyond the scope and intent of this book, a cursory discussion regarding evolvement of foundations is provided to aid in understanding of the impact of design on behavior. The primary focus is directed to foundation distress, cause, repair, and prevention. The discussion is generally limited to lightly loaded foundations constructed on expansive soils.

Various repair options are discussed and evaluated based on available documentation. Relative cost comparisons are provided, where possible. The question of whether or when to consider the need for remedial action as well as preventative measures is also covered. Estimating, structural inspections; factors to consider in selecting an inspector, engineer, or contractor; and case histories are also discussed. The material should prove invaluable to civil and structural engineers, geotechnicians, architects, contractors, developers, and serious engineering students as well as to realtors, lenders, appraisers, insurers, and homeowners.

Prerequisite areas such as site preparation, soil mechanics, foundation design, and behavior and chemistry of concrete are not included. This information can be found in any one or more of the references provided.

The selection of subject matter was influenced by other authors and my associates, to whom I am grateful. The extensive references will enable the serious student to explore the subjects in as much detail as he or she wishes.

I also wish to express appreciation to Brown Foundation Repair and Consulting, Inc., for making the publication possible and to my wife, Mozelle, and family for their patience and support in preparing the text, especially my son, Robert L. Brown, for his editorial assistance and my daughter, Candy, for the artwork.

Robert Wade Brown

INTRODUCTION

In areas with expansive soils and diverse climate, the property damage due to foundation failures exceeds the cost of all other natural calamities combined.[57] Figure I.1 is a map of the United States that shows the distribution of expansive clay. The darker shaded areas have expansive clay (generally montmorillonite) in relative abundance. Nine states can be included in the most affected group (Texas, Mississippi, Montana, Colorado, California, Wyoming, South Dakota, North Dakota, and Alabama). An additional eight states have similar conditions but to a lesser extent (Louisiana, Florida, Washington, Oregon, Nebraska, Arizona, Utah, and Tennessee). Twelve other states have limited problems, generally indicated by the lighter shading. These states have either a less-expansive clay (illite, for example) or montmorillonite in a much lower abundance. Other countries also suffer problems

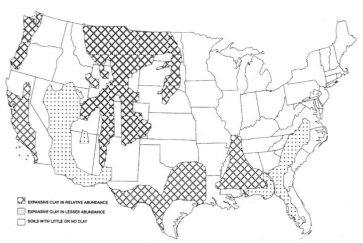

EXPANSIVE CLAY IN RELATIVE ABUNDANCE
EXPANSIVE CLAY IN LESSER ABUNDANCE
SOILS WITH LITTLE OR NO CLAY

FIGURE I.1 Distribution of expansive soils in the United States.[57]

in dealing with expansive clays. These include Australia, Canada, England, China, Mexico, the Netherlands, Japan, South Africa, and Israel.

Most expansive soils experience wide changes in volume upon access to various amounts of water. This accounts, in large part, for foundation failures that occur. An expansive soil generally swells upon access to water and shrinks upon loss of water.[15-17,26,69,75] Expansive soils are also referred to as cohesive.

As the soil loses moisture, it shrinks. Because most residential slab foundations are designed to be supported essentially 100 percent of their area, loss of support due to soil shrinkage can result in foundation settlement.[15-17,75] Pier and beam foundations respond similarily although they are affected less frequently. However, the consequences are normally neither too serious or too prevalent.[15,26] Soil shrink (settlement) can generally be arrested (and sometimes reversed) by the introduction of water.

As the soil swells, the volume increase is accompanied by tremendous potential forces. These have been reported to theoretically reach 20,000 lb/ft^2 (97,600 kg/m^2). This swell potential eclipses the structural weight of lightly loaded construction by several magnitudes. The resultant condition is referred to as *foundation heave* or *upheaval*. This is by far the most serious and most prevalent problem to which foundations are subjected.[15-17] Pier and beam foundations also suffer the consequences of heave but generally to a less dramatic extent and frequency. A relatively low increase in moisture can result in a rather substantial slab foundation heave of 1¼ in (3 cm) with a moisture change of only 4 percent.[102,103] Note, however, that the particular soil composition as well as the amount of swell potential produced by a given change in water percentage varies widely with the original in situ water content. (The cited soil had a plastic limit of 28 with an initial moisture of 20 percent.) A detailed discussion of Atterberg Limits (SL, PL, PI, LL, W percent) can be found in any text on soil mechanics.[15-17] A brief introduction, however, will follow.

Atterberg Limits were devised to help describe and explain the behavior of expansive soil upon impact with water. Figure I.2 presents a graphic representation of the Atterberg Limits. The liquid limit (LL) represents the moisture content at which the soil behavior changes from a plastic to a liquid state. The PL represent the point at which the soil moves from a semisolid state to a plastic state. The difference between the two is the plasticity index (PI) (i.e., LL − PL = PI). The greater the PI, the more expansive the soil. The shrinkage limit (SL) is the point separating the solid and semisolid states. At this level, the soil can no longer experience a decrease in volume upon extraction of water. And finally, the water content (W percent) is the natural (or in situ) water content of the soil. Selective use of these values is a useful tool for the prediction of soil behavior—particularly when dealing with expansive soils. Other properties such as unconfined compressive strength, shear strength, density, particle sizes, swell and shrink potentials, and chemical composition are also factors of concern. Nonexpansive soils can also experience foundation problems. For example, soil "settlement" can be the result of some variety of

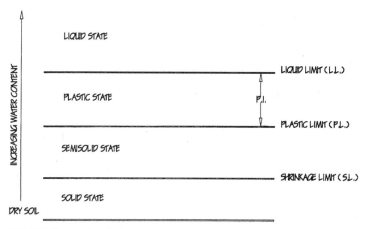

FIGURE I.2 Atterberg limits and related indices.

(1) lateral displacement, (2) compaction of fill, or (3) collapse of internal voids. Lateral movements can be the result of soils sliding or sloughing. This is generally associated with construction on slopes with unstable bearing soils. Movement is often precipitated or exacerbated by the intrusion of water into the soil to the extent that both cohesion and structural strength are threatened or destroyed. California mud slides are one prime, though extreme, example. The structural damage caused by these problems is often complete destruction. When remediation is possible, the situation is addressed in a manner generally consistent with that for acute settlement. Repair of severe lateral movement is generally beyond the methods available to the repair contractor. However, the contractor can often provide measures to stop lateral movement.

Settlement can also occur as a result of consolidation, or compaction, of fill, base, or subbase materials. With respect to residential construction, the most common problem deals with construction on either abnormally thick granular or a sanitary fill. In either case, over time, the intended bearing soils fail due to consolidation. Normal settlement of fill is often active for periods up to 10 years—somewhat dependant upon the cycles of precipitation and drought. Sanitary fills can also be active for long periods of time due to voids continually provided by the decay of organic materials.

Other causes of settlement are a result of either so-called collapsing soils or the collapse of soil structure into internal voids, often the result of decaying organic material.

In many cases, deep grouting is a required remedial procedure for the correction of either problem. Once the deep-seated cause has been addressed, procedures common to foundation settlement can be used to "relevel" the structure.

Under appropriate conditions, specific soils, referred to as collapsing soils, are susceptible to a substantial reduction in void ratio upon addition of water. These soils might exist naturally at low moisture content with apparent high strengths largely attributable to particle bonding. Water destroys this structural bond and allows local compression. The majority of soils that tend to experience this collapse are aeolian (wind) deposits such as sands, silts, sandy silts, or clayey sands of low plasticity in which there is a loose structure with relatively low densities. As opposed to time-dependent consolidation (reduction of void ratio permitted by removal of pore water), settlement of these collapsing soils occurs somewhat rapidly upon the influx of water.[30,33,55,56,67,107] The distribution of major loess deposits within the United States is shown in Figure I.3.

Upheaval can be result of frost heave. In frigid climates residual soil moisture can freeze, expand, and distort the foundation. Refer also to Sec. 1.3.4.

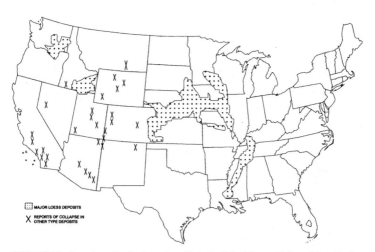

FIGURE I.3   Location of major loess deposits in the United States. (*Adapted from Dudley, 1970; used with permission of ASCE.*)

FOUNDATION
REPAIR
MANUAL

CHAPTER 1
HISTORY

1.1 INTRODUCTION

A foundation is that part of a structure that is in direct contact with the ground and that transmits the load of the structure to the ground. This load, W_T, is the sum of live loads (W_L) and dead loads (W_D). The dead load is the weight of the empty structure. The live load is the weight of the building contents plus wind, snow, and possibly earthquake forces, where applicable. The magnitude of W_T is often assumed to be in the range of 50 to 150 lb/ft^2 (2.4 to 7.2 kN/m^2) for residential construction of no more than two stories. In structures with a typical gabled roof, approximately 70 to 80 percent of the total loaded (W_T) is transferred to the perimeter beam. Often this perimeter load is estimated to be in the range of 400 to 600 lb. (19 to 29 kN/m^2) per linear foot of 12-in (30-cm) -wide perimeter beam. This load must be supported by the soil. The characteristic of the soil that measures its capacity to carry the load is the unconfined compressive strength q_u. The actual value of q_u is determined by laboratory tests. As discussed in the following paragraphs, other factors, such as climatic condition and bearing soil characteristics, also influence foundation design and selection.

There are two general categories for residential foundations: pier and beam, and slab. (In some areas, a third type is recognized that utilizes piles for the basic support.)

1.2 PIER AND BEAM CONSTRUCTION

The first foundation was quite simplistic and without interior floors. In the seventeenth century the "comforts" of home were enhanced (Fig. 1.1a).[94]

The foundation has rough-hewn timbers supported at the corners by selected native stones. Cross timbers were added that supported flooring. Initially the living space was limited to 20 by 30 ft (0.6 to 0.9 m), largely because longer, straight logs were hard to come by. As time passed the living area was increased by either adding "cells" or a second story. In prenineteenth century England, the "Victorian" foundation was popular.[41] This is illustrated as Fig. 1.1b.

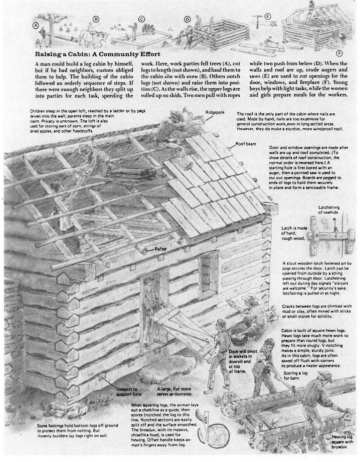

Raising a Cabin: A Community Effort

A man could build a log cabin by himself, but if he had neighbors, custom obliged them to help. The building of the cabin followed an orderly sequence of steps. If there were enough neighbors they split up into parties for each task, speeding the work. Here, work parties fell trees (A), cut logs to length (not shown), and haul them to the cabin site with oxen (B). Others notch logs (not shown) and raise them into position (C). As the walls rise, the upper logs are rolled up on skids. Two men pull with ropes while two push from below (D). When the walls and roof are up, crude augers and saws (E) are used to cut openings for the door, windows, and fireplace (F). Young boys help with light tasks, while the women and girls prepare meals for the workers.

Children sleep in the open loft, reached by a ladder or by pegs driven into the wall; parents sleep in the main room. Privacy is unknown. The loft is also used for storing ears of corn, strings of dried apples, and other foodstuffs.

Ridgepole

The roof is the only part of the cabin where nails are used. Made by hand, nails are too expensive for general construction work, even in long settled areas. However, they do make a sturdier, more windproof roof.

Roof beam

Door and window openings are made after walls are up and roof completed. (To show details of roof construction, the normal order is reversed here.) A starting hole is first bored with an auger, then a pointed saw is used to cut out openings. Boards are pegged to ends of logs to hold them securely in place and form a serviceable frame.

Rafter

Latchstring of rawhide

Latch is made of hard, tough wood.

A stout wooden latch fastened on by pegs secures the door. Latch can be opened from outside by a string passing through door. Latchstring left out during day signals "visitors are welcome." For security's sake, latchstring is pulled in at night.

Cracks between logs are chinked with mud or clay, often mixed with sticks or small stones for solidity.

Cabin is built of square hewn logs. Hewn logs take much more work to prepare than round logs, but they fit more snugly. V-notching makes a simple, sturdy joint. As in this cabin, logs are often sawed off flush with corners to produce a neater appearance.

Scoring a log for barn

Door will pivot in sockets in doorsill and at top of frame.

Sleepers to support floor

A large, flat stone serves as doorstep.

When squaring logs, the axman lays out a chalkline as a guide, then scores (notches) the log to this line. Notched sections are easily split off and the surface smoothed. The broadax, with its massive, chisellike head, is used for hewing. Offset handle keeps axman's fingers away from log.

Stone footings hold bottom logs off ground to protect them from rotting. But slovenly builders lay logs right on soil.

Hewing log square with broadax

FIGURE 1.1a   Seventeenth century foundation—old west.

Shelter for a Frontier Family

The log cabin was introduced into America by Swedish and Finnish colonists on the Delaware in the 1630's and spread later by the Germans and Scotch-Irish. Predominantly a feature of the backcountry, it really came into its own with the great migrations across the Appalachians after the Revolution. The advantage of the log cabin was that it could be built quickly, needed few tools—in a pinch, an ax alone would do—and the building material grew on the site. Never intended as a permanent dwelling, it provided a durable shelter until a family could afford a larger, more comfortable house of framed lumber, brick, or stone. The old log cabin was often built thriftily into the new house or turned into a shed. The one shown below in the process of construction is based on the description of an actual cabin built in Illinois in 1820 for a well-to-do family. The log cabin became a sentimental symbol of pioneer days during the presidential campaign of 1840, when William Henry Harrison's supporters claimed that he would be happy in a log cabin, like a man of the people.

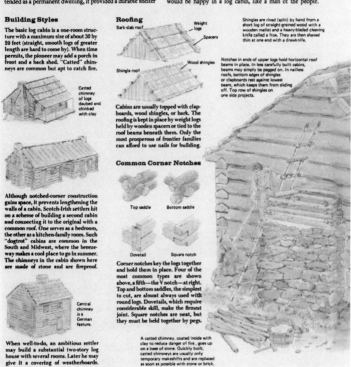

Building Styles

The basic log cabin is a one-room structure with a maximum size of about 30 by 20 feet (straight, smooth logs of greater length are hard to come by). When time permits, the pioneer may add a porch in front and a back shed. "Catted" chimneys are common but apt to catch fire.

Catted chimney of logs daubed and chinked with clay

Although notched-corner construction gains space, it prevents lengthening the walls of a cabin. Scotch-Irish settlers hit on a scheme of building a second cabin and connecting it to the original with a common roof. One serves as a bedroom, the other as a kitchen-family room. Such "dogtrot" cabins are common in the South and Midwest, where the breezeway makes a cool place to go in summer. The chimneys in the cabin shown here are made of stone and are fireproof.

Central chimney is a German feature.

When well-to-do, an ambitious settler may build a substantial two-story log house with several rooms. Later he may give it a covering of weatherboards.

Roofing

Bark-slab roof
Weight logs
Spacers

Wood shingles

Shingle roof

Cabins are usually topped with clapboards, wood shingles, or bark. The roofing is kept in place by weight logs held by wooden spacers or tied to the roof beams beneath them. Only the most prosperous of frontier families can afford to use nails for building.

Common Corner Notches

Top saddle
Bottom saddle

Dovetail
Square notch

Corner notches key the logs together and hold them in place. Four of the most common types are shown above, a fifth—the V notch—at right. Top and bottom saddles, the simplest to cut, are almost always used with round logs. Dovetails, which require considerable skill, make the firmest joint. Square notches are neat, but they must be held together by pegs.

Shingles are rived (split) by hand from a short log of straight-grained wood with a wooden mallet and a heavy-bladed cleaving knife called a froe. They are then shaved thin at one end with a drawknife.

Notches in ends of upper logs hold horizontal roof beams in place. In less carefully built cabins, beams may simply be pegged on. In nailless roofs, bottom edges of shingles or clapboards rest against lowest beam, which keeps them from sliding off. Top row of shingles on one side projects.

A catted chimney, coated inside with clay to reduce danger of fire, goes up on a base of stone. Quickly built, catted chimneys are usually only temporary makeshifts and are replaced as soon as possible with stone or brick.

FIGURE 1.1a (*Continued*) Seventeenth century foundation—old west.

The early twentieth century foundation, generally recognized for residential construction, was a pier and beam. The "primitive" version involved nothing more than the use of wood "stumps" placed directly on the soil for structural support. These members, usually cedar or bois d'arc, support the sole plate as well as the interior joists or girders. Recent variations are more sophisticated (Fig. 1.2).[93] Figure 1.2a shows one design for a wood perimeter foundation. In this instance, the foundation extends below grade to accommodate a frost line.

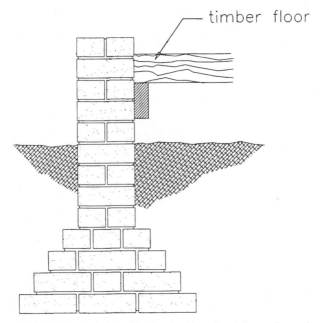

FIGURE 1.1b Typical English Victorian House foundation prenineteenth century.[41]

Figure 1.2b depicts an interior floor support. The wood interior (or steel) post is supported on a concrete footing. In earlier versions, the support might have rested directly on the ground surface. (Dr. Sazinski, regional engineer for HUD also presents information on wood foundations.[17]) This design then evolved to the use of a continuous concrete beam to support the perimeter loads. As a rule, the same wood piers initially served to support the interior. This variation was precipitated largely by the advent of brick or stone veneer. Further changes in construction requirements, including the awareness of problems brought about by unstable soils, encouraged the addition of concrete piers to support the perimeter beam and concrete piers and pier caps; the latter provided better support for the interior floors (girders) (Fig. 1.3).

As opposed to the slab, the design of pier and beam foundations has not relegated itself to system of "cookbook" standardization. This design requires a separate system of engineered steel-reinforced concrete pier and beams. The exact specifications are dictated by such factors as structural load, soil and site conditions, and subsurface water.

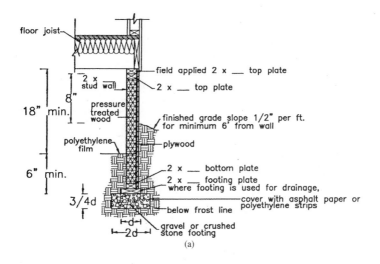

floor joist

field applied 2 x ___ top plate
2 x ___ top plate

2 x
stud wall

8"

18" min.

pressure
treated
wood

finished grade slope 1/2" per ft.
for minimum 6' from wall

polyethylene
film

plywood

6" min.

2 x ___ bottom plate
2 x ___ footing plate
where footing is used for drainage,
cover with asphalt paper or
polyethylene strips
below frost line

3/4d

gravel or crushed
stone footing

d

2d

(a)

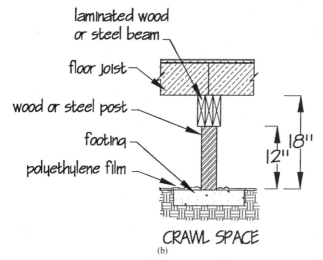

laminated wood
or steel beam

floor joist

wood or steel post

footing

polyethylene film

18"

12"

CRAWL SPACE
(b)

FIGURE 1.2 (*a*) Crawl space detail; (*b*) corner post detail.[93]

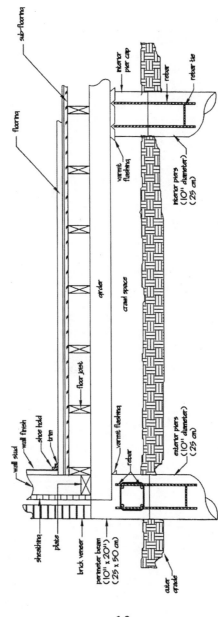

FIGURE 1.3 Twentieth century typical pier and beam.

In the following discussion, pier and beam construction is that design whereby the perimeter loads are carried on a continuous beam supported by piers drilled into the ground, presumably to a competent bearing soil or stratum. The interior floors are supported by piers and pier caps that sustain the girder and joist system of a wood substructure. All foundation components are assumed to be steel-reinforced concrete. For practical purposes, there are two principal variations of the pier and beam design: the normal design, in which the crawl space is on a grade equivalent to the exterior landscape, and the relatively new "low profile," in which the crawl space is substantially lower than the exterior grade (Figs. 1.3 and 1.4).

The stability of the pier and beam design depends, to a large extent, upon the bearing capacity and nature of the soil at the base of the piers. Assuming ideal conditions, this design provides a stable foundation. Difficulties can occur if a stable bearing material is not accessible as a pier base. In some instances, piers are belled to spread the structure load and allow the use of less competent soils to support the weight. Belled piers might also provide an adequate resistance to upheaval in expansive soils. This may not always offer a foolproof solution. The belled area tends to anchor the pier and integral beam. However, should sufficient upward force (upheaval) be transmitted to the beam, a cracked foundation could result. The use of void boxes under the beams can provide some limited protection against upheaval stress. However, in many instances the precaution fails because either the soil swell exceeds the void provided by the boxes or the void simply fills in with migrating soil. Proper moisture control still has the most successful stabilizing effect. Effective exterior grade, adequate watering, and sufficient ventilation are all forms of proper moisture control.

The low-profile design (Fig. 1.4) presents inherent problems relative more to excessive moisture than to actual foundation distress. The low-profile design encourages the accumulation of water in the crawl space beneath the floors. This tends to result in mold, mildew, rot, termites, and, if neglected, differential foundation movement. Proper ventilation and exterior grade will normally arrest any serious structural distress. In persistent problems of moisture accumulation, forced-air blowers to provide added ventilation and/or chemical treatment to kill the mildew can be beneficial. The practice of covering the crawl space with polyethylene sheets should be avoided. Although the film reduces the humidity in the crawl space, it also develops a "terrarium," with subsequent threat to the foundation. Moisture accumulation can be best handled by other means, as noted. The air vents should never exist below the exterior grade; they would channel surface water directly into the crawl space.

In certain areas that have soil problems more serious than those experienced in the United States, more intricate foundation design is suggested. For

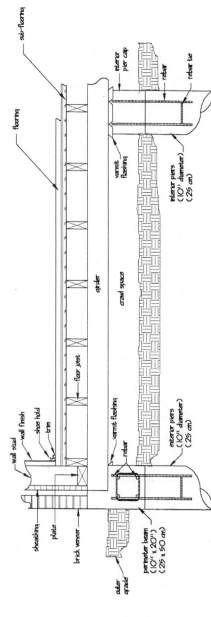

FIGURE 1.4 Typical low-profile pier and beam.

1.8

example, in Adelaide, Australia,[52] where plasticity indexes (PIs) often fall in the range of 60 or more, "pier and beam" designs illustrated by Fig. 1.5 are used. Note that both examples lack the conventional piers associated with the pier and beam depicted in Figs. 1.3 and 1.4. (Actually, this design is not typical of the U.S. pier and beam; it is referred to as such because the interior floors do not rest on bearing soil.)

1.3 SLAB

Slab construction of one variety or another probably constitutes the majority of new construction in geographic areas exposed to high-clay soils. The concrete slab foundation first appeared in the late 1950s. Until 1968, construction was rather haphazard, with no concentrated effort to either standardize it or provide specific engineering designs. In 1968, the BRAB Bulletin was published.[36] This publication provides specific design parameters for steel-reinforced concrete slab on ground foundations to meet varying requirements of climates, structural load, soil, and site conditions.

1.3.1 Conventional

Slab foundations come in various configurations as determined by such factors as soil conditions, load considerations, and the variable nature of weather (Figs. 1.6 to 1.9). As a rough rule of thumb, soil expansion greater than 2 to 3 percent is considered potentially dangerous and requires special design considerations.

In most cases, the structural load is designed to be transmitted essentially 100 percent to the bearing soil. As a place to start, the FHA (or HUD) designs pose a handy reference. Most of the FHA material was obtained from independent research. Fortunately, many of the data are available to the industry. One example is the much referenced publication *Criteria for Selection and Design of Residential Slab on Ground,* BRAB Publication No. 33.[36] Many other publications have been sponsored by the government, but as far as references can determine, none have substantially changed the principal design criteria postulated by the BRAB publication.

Another factor that suggests special concern is any anticipated condition wherein the foundation could be expected to act as a bridging member. The BRAB book addresses this as a support factor C, which accommodates minimal cantilever conditions. Ideally, C would be 1.0, indicating 100 percent support by the soil. As the support factor deviates below 1.0, indicating less than 100 percent support by the soil, special design considerations are demanded.

Figures 1.6 to 1.9 depict typical representations of various slab designs—based to a large extent on the FHA design. Because structural

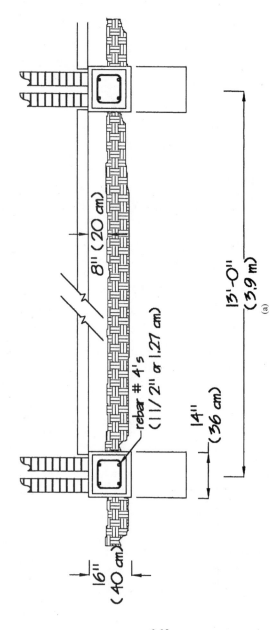

FIGURE 1.5 (a) Soil with a plasticity index (PI) less than 10.

8" (20 cm)

13'-0" (3.9 m)

(a)

rebar # 4's (11/2" or 1.27 cm)

14" (36 cm)

16" (40 cm)

1.10

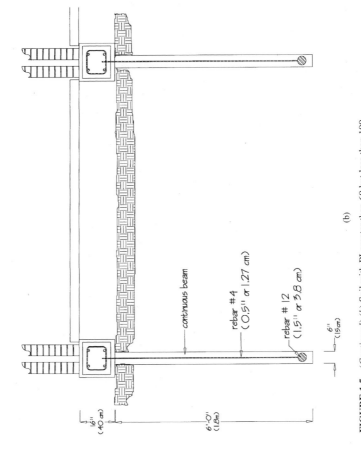

FIGURE 1.5 (*Continued*) (*b*) Soil with PI greater than 60 but less than 100.

continuous beam

rebar #4
(0.5″ or 1.27 cm)

rebar #12
(1.5″ or 3.8 cm)

6″
(15cm)

16″
(40 cm)

6′-0″
(1.8m)

(b)

1.11

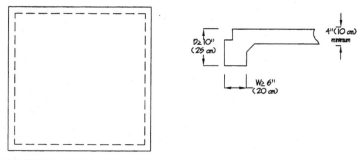

FIGURE 1.6 Typical FHA type I slab.

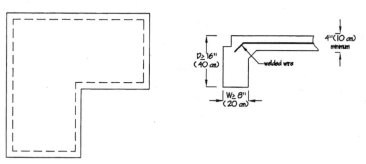

FIGURE 1.7 Typical FHA type II slab.

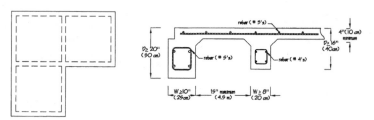

FIGURE 1.8 Typical FHA type III slab.

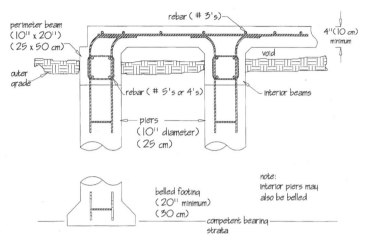

FIGURE 1.9 Typical FHA type IV slab.

design is basically beyond the scope of this book, the primary concern will be major differences in design that materially influence foundation failure or repair. Bear in mind that resistance of a beam to differential deflection is influenced more by depth than by width. In fact, other things being equal, a twofold increase in depth will improve the resistance to deflection by a factor of 8. Table 1.1 presents some of the major differences in soil (based on classification) and climatic conditions that combine to influence the acceptable foundation design. The symbol W_T is used to designate the total unit foundation load, dead load plus live load ($W_D + W_L$), and q_u represents the unconfined compressive strength of the soil. Note the column for slab type. This provides the generally preferred slab for the various conditions of soil and climatic conditions.

Note in particular the dependence of the design upon the plasticity index PI and climatic conditions C_W. The PI is a dimensionless constant that has a direct relation to the affinity of the bearing soil for volumetric changes with respect to moisture variations. (Refer to the Introduction for a prelude to the Atterberg Limits.) The higher the value of the PI, the greater the volatility of the soil. Generally, the volumetric changes are directly related to specific clay content of the soil as well as to differential moisture. [For example, Tucker and Poor[102] reported in their study that a slab foundation (typical FHA II) constructed on a soil with a PI of 42 experienced a vertical deflection of about $1^1/4$ in (3.2 cm) upon a moisture content increase of 4 percent.]

The liquid limit (LL) is determined by measuring the water and the number of blows required to close a specific-width groove of specified length in

TABLE 1.1 Soil Differences and Climate Conditions that Influence Foundation Design

Soil type	Minimum densities, q_u	Climatic rating, C_W	Slab type	Reinforcement
Gravel	All densities	All	I*	None except wire mesh at openings, step-downs, etc.
Gravel sands, low PI silts, and clays	Dense	All	I	Same
Same	Loose (non-compacted)	All	II†	Lightly reinforced wire mesh, rebar in perimeter beam
Organic or inorganic clay and silts	PI<15 and $q_u/W \geq 7.5$	All	II	Same
Same	PI > 15	$C_W > 45$	II	Same
Same	PI > 15 q_u/W_T between 2.5 and 7.5	$C_W < 45$	III	Rebar plus interior beams
Same	q_u/W_T between 2.5 and 7.5	All	III	Same
Same	$q_u/W_T < 2.5$	All	IV	Structural slab (not supported directly on soil)

*Maximum partition load 500 lb/ft (757 kg/m).

†Maximum partition load 500 lb/ft unless loads are transmitted to independent footings.
Source: Simplified from BRAB Report 33.[36]

a standard liquid limit device. The PL is determined by measuring the water content of the soil when threads of the soil $^1/_8$ in (0.3175 cm) in diameter begin to crumble.[16,17,65] The PI is the difference between the two (LL − PL = PI). Finally, q_u is the unconfined compressive strength of the proposed bearing soil, determined experimentally.[16,17,65]

Again from the BRAB Report, only the top 15 ft of bearing soil influences foundation stability. The top 5 ft carries 50 percent of the imposed load; the second 5 ft, 33 percent; and the final 5 ft, about 17 percent. Other research presented throughout this book will suggest that the concern for soil behavior to the depth of 15 ft is probably excessive (refer to Chap. 8). Along this line of thought, Fua Chen's paper, "Practical Approach on Heave Prediction," questions the reliability of theoretical approaches to predict heave.[27] The heave prediction methods are all based on the assumed depth of wetting, which varies considerably among various investigators. He suggests that heave predictions are generally much greater than those actually measured in the field. Does this mean that the seasonal depths of soil moisture change are, in fact, considerably less than values usually assumed?

Figure 1.10 shows the climatic ratings (C_w) for the United States issued by the U.S. Department of Commerce. (The Thornwaite Index is another rating system that shows the impact of climate on the performance of structures on expansive soils.[13] Refer to the Glossary for a definition.) As a matter of interest, the British Standards Institute, "Code for Foundations," CP2004:1972,[10] states that seasonal variations in their heavy clay soils occur to a depth of 5 to 6 ft (1.5 to 2.0 m) under areas not watered. They further suggest that foundations for traditional masonry structures be founded at a depth of at least 3 ft (1 m) to reduce relative movement to acceptable limits. Their heavy clays are quite similar to those of the United States.

In essence, as the design and environmental conditions become more severe, the slab must be strengthened. This is accomplished by increasing the size and the frequency of both the beams and reinforcing steel.

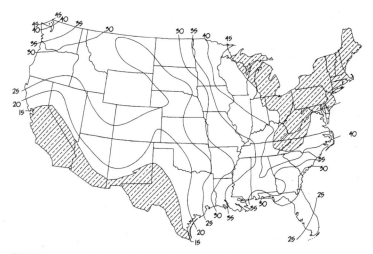

FIGURE 1.10   Climatic ratings for continental United States.

1.3.2 Post Tension

A relatively recent innovation is the use of post-tension stress to the slab. Concrete is adequate in compression but has limited tensile strength. The cables are intended to enhance the tensile strength of the slab through the compressive effects of the cable tension. So it is in theory; in practice, the results do not always meet expectations. Perhaps the problem lies with the fact that, in field performance, the compressive loads are not always perpendicular to the stress, and vectors are created that tend to destroy the benefits of the principle. (In simpler terms, adequate attention is not given to keep the cables straight, parallel, and/or perpendicular.) The author's experiences indicate that failures with post-tension residential slabs are more preponderant, on a percentage basis, than with conventional deformed-bar slabs. This same observation has been documented by others.[2] As indicated above, many of these problems are probably caused by poor field practices and, as such, may not reflect a design defect.[2] Figure 1.11 illustrates a typical post-tension slab. As a rule, the elongation of the cable is intended to be 1 in (2.54 cm) per 12-ft (4-m) -length of tendon. For more information on post-tension foundations, refer to *Design and Construction of Post Tension Slabs-on-Grade*.[89] The cables normally consist of a $^3/_8$- to $^1/_2$-in (0.95- to 1.27-cm) -diameter seven-wire strand having a minimum ultimate tensile

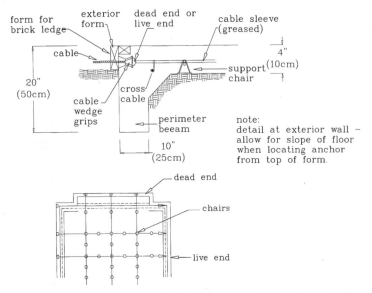

FIGURE 1.11   Typical post-tension slab (for detail purposes, slab is not drawn to scale).

strength of 270,000 lb/in^2 (186 kN/cm^2). Stressing normally occurs between 72 h (optimum) and 14 days (maximum) after pouring the concrete. The cables are stressed to 80 percent of ultimate design strength, approximately 2300 lb/in^2 (1.58 kN/cm^2) for ^{3}/8-in (0.95-cm) cables and 4200 lb/in^2 (2.89 kN/cm^2) for ^{1}/2-in (1.27-cm) cables.

An issue not addressed in Fig. 1.11 is the use of a plastic membrane between the bearing surface and poured concrete. The plastic membrane is intended to serve as both a moisture barrier and, to a lesser extent, a friction reducer between the poured concrete and its bearing surface. As far as residential and light commercial construction is concerned, the reduction in the effective coefficient of friction probably has more significance to control movements due to concrete shrinkage and temperature variations than to post tensioning. This is because, first, the total slab movement due to post tensioning is quite small, usually less than ^{1}/16 in, and second, the presence of stiffening beams further retards any tendency for movement. Refer again to the *Design and Construction of Post Tension Slabs-on-Grade*[89] for more detail.

1.3.3 Design Variations

Special construction, and/or site conditions, suggest various changes or considerations for foundation design. For example, in areas with low or moderately expansive soil, often of low or noncohesive nature, a typical design might be similar to Fig. 1.12. Areas representing this criteria would include parts of Florida, Oklahoma, and other states.

A somewhat similar design is used in the United Kingdom, where the soil is both expansive and cohesive (Fig. 1.13*a* and *b*).[41] The montmorillonite content averages 20 to 40 percent, which is roughly equivalent to that exhibited by the more volatile U.S. soils. The climate in the United Kingdom is such that soil moisture remains both high and relatively constant year-round. Typically, the annual rainfall in London is 30 in (76 cm), which occurs over 152 days. (By comparison, about 70 percent of the typical rainfall in Texas occurs in 15 days).

Typically, a London clay might have a PL = 26, LL = 69, PI = 43, and a moisture content of 30 percent. With an initial, reasonably consistent moisture content of this magnitude, little swell potential remains; however, a loss of soil moisture could create substantial subsidence. Tomlinson suggests that a minimum foundation depth to guard against seasonal moisture variations would be 3 ft (0.9 m).[101] This assumes a residential foundation with bearing loads of 1.0 ton/ft (3030 kg/m) of load bearing wall (perimeter) on a stiff fissured clay.[7]

Another variation from the U.S. slab foundation is represented by the "raft" design common to parts of Australia.[52,53] Figure 1.14 shows such a foundation. The top view is comparable to Fig. 1.8, except the width of the beam is less. A typical Adelaide soil might have a PL = 30, LL = 90, PI = 60+, and a moisture content of 15 to 35 percent. The montmorillonite content can

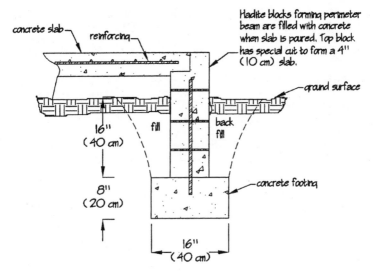

FIGURE 1.12 Variation: Slab foundation (low PI).

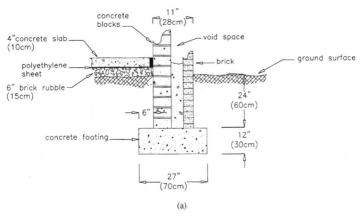

(a)

FIGURE 1.13 (a) Variation: Slab foundation in United Kingdom (typical).

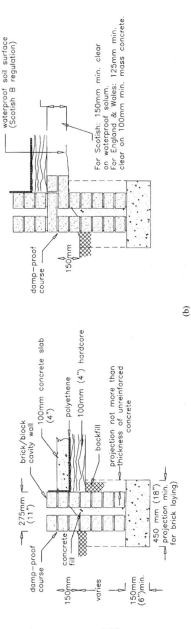

waterproof soil surface
(Scotish B regulation)

damp-proof
course

150mm

For Scotish: 150mm min. clear
on waterproof solum.
For England & Wales: 125mm min.
clear on 100mm min. mass concrete.

(b)

275mm (11")

brick/block
cavity wall

100mm concrete slab
(4")

polyethene

100mm (4") hardcore

backfill

projection not more than
thickness of unreinforced
concrete

damp-proof
course

concrete
fill

150mm

varies

150mm
(6")min.

450 mm (18")
projection min.
for brick laying

FIGURE 1.13 (*Continued*) (*b*) Typical strip footing with ground-bearing floor slab (left) and a suspended timber floor (right).

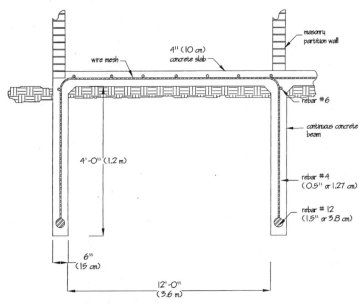

FIGURE 1.14   Variation: Slab foundation, Australia.

average something like 70 percent. In areas with PI of 10 or less, the foundation design is almost identical to that for the monolithic U.S. slabs.

1.3.4   Frost Heave

Expansive soils aren't the only serious concern facing foundation design. For example, building on permafrost can also represent some concern. Here the problem is to erect and occupy heated structures without thawing the permafrost. Should the ice-rich permafrost melt, building stability is sacrificed due to excessive settlement.[4]

In Canada, the current standard is to elevate the residential structure on piles, thus allowing air circulation beneath the building to remove heat leaked through the floor and preventing the melt of the permafrost. The primary disadvantage to this approach is cost brought about by transportation of the piles, installation of piles in frozen ground, and design and construction problems inherent to structural floors. A recent program (1987) jointly sponsored by the Institute for Research in Construction, National Research Council of Canada, and Yukon Government tested a new innovation—a heat pump chilled foundation.[4] The insulated slab on grade foundation, heat pump system, and in-ground heat exchangers were designed to maintain the permafrost (prevent melting) beneath the heated structure (Fig. 1.15). The insulation was designed to prevent a thaw in the ice-rich subgrade for 8 to 9 months should a heat pump fail. The heat exchangers are placed in a sand layer within the granular fill used to provide a level building site. The heat exchangers are made of high-density polyethylene filled with methyl hydrate solution and laid out in a serpentine pattern. The concept resembles a refrigerator acting in reverse. Heat flow escaping through the floor is captured by heat exchangers and pumped back into the building to supplement heating requirements. The ground is chilled, whereas the building is heated. Time and future monitoring will decide the future for this rather unique design.

Foundation designs are numerous. However, the foregoing gives a brief glimpse at some of the more interesting accepted practices.

1.4 BASEMENTS

Generally, in present times, the use of basements as part of the foundation design is limited to areas where either real estate costs are quite high or a deep frost line exists. The reason for this trend is simple. Low-cost real estate makes it much less expensive to spread out than to excavate. A deep frost line

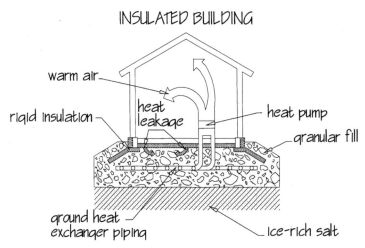

FIGURE 1.15 Foundation on permafrost.

often makes the basement route more viable from a construction cost view. Figure 1.16 presents a typical basement-foundation design. In the past, when coal furnaces were more prevalent, at least partial basements were common in areas not inclusive in the land cost–frost line criterion. In that event, the furnaces and coal storage were situated below the floors and necessitated some basement.

1.5 UTILITY CONCERNS TO FOUNDATION REPAIR

Concern is often given to the problems that might be incurred by utility lines during any foundation-leveling procedure. The utility lines in question would be (*a*) electricity, (*b*) telephone, (*c*) gas, (*d*) water, and (*e*) sewer or septic.

1.5.1 Electricity

Electric service is most often overhead and generally not much of a concern. In rare occasions where the structure is to be raised 1 ft (0.3 m) or more, it would be wise to cut the service off at the pole. When electric lines are sub-

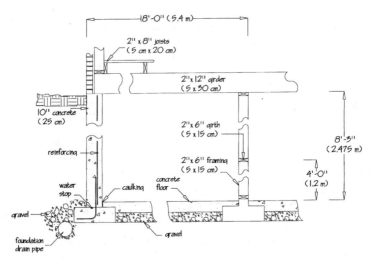

FIGURE 1.16 Basement construction.

surface, the utility company should be notified prior to any excavation work and asked to tag the lines.

1.5.2 Telephone

Concerns for telephone service are the same as for electricity.

1.5.3 Gas

Gas service should never be run under a slab foundation. The lines are underground up to the perimeter of foundation and then run overhead. The pier and beam foundation may have gas lines underneath in the crawl space. In this location the lines are readily visible and easy to avoid. The only two real concerns with gas lines might be encountered when large raises are required or where the underground lines don't follow the "obvious" path. In either event, the best precaution is to ask the gas company to tag the lines and/or turn the gas supply off. Gas immediately into the house can be controlled at the meter. Gas lines on the street (supply) side of the meter must be shut off by the utility company.

1.5.4 Water

Water lines frequently follow the same path as gas lines and pose similar infrequent problems. Pier and beam foundations seldom, if ever, suffer any consequences. Slabs, generally, enjoy the same immunity except in rare instances. Occasionally, a "misplaced" water line (copper tube) might be severed by a drill bit preparatory to mudjacking. This occurs probably less than once in 600 jobs (0.16 percent).

1.5.5 Sewer (septic)

Sewer lines are the most frequent "casualty" of foundation repair. Generally, pier and beam foundations do not offer particular concern unless the amount of raise is excessive [i.e., more than about 6 in (15 cm)] and causes a line separation. In the case of slab foundation the concern gets somewhat more serious. Sewer lines can occasionally be damaged either at the perimeter during excavation or in the interior generally during drilling preparatory to mudjacking. The interior lateral lines sometimes fill with grout during mudjacking. When this occurs, it is usually the result of an already damaged pipe. (The grout itself is not likely to create any damage to the system.) To combat the problem of damage to underground sewer lines two precautions are utilized: (1) Care is taken not to drill in the immediate vicinity of "suspected" pipe locations, and (2) water is run continuously during pumping in areas with plumbing. The water elutriates the cement from the grout, leaving a "pile of dirt" that has no tendency to set up. Alternatively, at the first sign of any stoppage, pumping ceases, the owner is notified, and a "roto-rooter" company is called in. Because the usual grout requires several hours to attain initial set, the blockage is easily removed at little cost. Grout into the sewer system is quite infrequent, less than 1 to 2 percent. Drilling into a sewer line occurs at about the same frequency, and most of this results from a "misplaced" sewer line. The bottom line is if you have a foundation problem, the rest of the concerns are inconsequential. Most foundation problems, relative to sewer line separation, occur prior to any foundation repairs.

CHAPTER 2
SUPPORTING INTERIOR FLOORS

2.1 INTRODUCTION

Foundation repairs are generally categorized by cause—settlement or upheaval. Pier and beam foundations, generally, are more susceptible to settlement problems. Between the two foundation problems, upheaval is by far the most prevalent and the more costly.[15–17,26]

Regardless of the cause of failure, the approaches to normal repair are quite similar. Whether a perimeter has settled or the interior heaved, the normal repair procedure is to underpin (raise) the perimeter. (Perimeter upheaval, though rare, can be lowered.[15–17]) Interior floors are then "leveled" by shimming the interior pier caps or in the case of slabs, mudjacking. Underpinning is discussed in Chap. 5 and mudjacking, in Chap. 3.

2.2 SHIMMING EXISTING CONCRETE PIER CAPS

Leveling interior floors by shimming on *existing concrete pier caps* is the simplest of all foundation repair. This is a problem characteristic to pier and beam foundations. Several times during the life of the residence the need to level interior floors will arise. This can be considered as "routine maintenance" much the same as repainting wood surfaces. If the existing piers are sound and properly located and assuming access, shimming requires nothing more than raising the girders to the desired elevation and placing wedges or shims on top of the pier caps to retain the position (Figs. 1.3 and 1.4). Dimension hardwood or steel are used for major gaps. Cedar shingles are often used to do fine grading.[15–17] Some people are confused regarding the advisability of using shingles. The concern is not founded in fact. True, the shingle will compress under the load of the structure. However, this compression will continue only to the point where differences in densities are met. This compression is taken into account by the contractor. Without the use of shingles, fine grading [less than $1/4$ to $3/8$ in (0.6 to 0.95 cm)] is not as practical.[16] Also bear in mind that shingles have been used in new construction for over 50 years to level plates, frames, and girders. The use of thin steel shims can be a viable option.

2.3 SHIMMING USING SUPPLEMENTAL PIER CAPS

When existing floor supports are deficient for one reason or another, the need for new pier caps arises. The ideal solution would be the installation of "new" drilled piers as shown in Fig. 5.1. However, this approach is prohibitively expensive for existing structures because of restricted access. The flooring and perhaps partition walls and joist would have to be removed to provide access. Thus, supplementary supports normally consist of precast concrete pads (leveled into the soil surface) with masonry blocks (usually Hadite) serving as a pier cap (Fig. 2.1).

The support depicted in Fig. 2.1 involves a concrete base pad, a concrete pier cap, suitable hardwood spacers, and a tapered shim for final adjustment. The base pad can be either poured in place or precast. The choice and size depends largely upon the anticipated load and accessibility. For *single-story frame* construction, the pad is normally precast, at least $18 \times 18 \times 4$ in $(46 \times 46 \times 10$ cm) thick, with or without steel reinforcing.

For *single-story and normal two-story brick-construction* load conditions, the pad should be steel-reinforced and at least $24 \times 24 \times 4$ in $(61 \times 61 \times 10$ cm) thick. For unusually heavy load areas (e.g., a multiple-story stairwell), the pad should be larger, thicker, and reinforced with more steel. In the latter case, the pad is normally poured in place. (The added weight creates severe handling problems for precast pads.) In any event, the pad is leveled on or into the soil surface to produce a solid bearing. Conditions rarely warrant any attempt to place the pads materially below grade. Often these may be 3 ft $\times$ 3 ft $\times$ 6 in $(0.9 \times 0.9 \times 0.15$ m) thick and reinforced with two mats of #3 rebar. Once the pad is prepared, the pier cap can be poured in place or precast; in most cases, the limited work space favors the precast cap. Choices for the precast design include concrete cylinders, Hadite blocks (lightweight concrete), or other square masonry blocks. (Ideally, the head of the pier cap

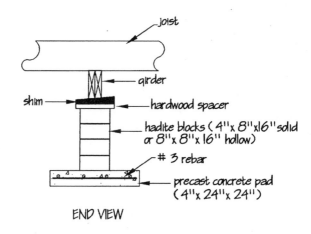

END VIEW

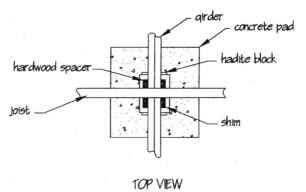

TOP VIEW

FIGURE 2.1 Supplemental interior pier cap (typical).

should be at least as wide as the girder to be supported.) Either material is acceptable for selected loads. The Hadite blocks, for example, may occasionally be inadequate for multiple-story concentrated loads unless the voids within the blocks are filled with concrete, or solid blocks are utilized. [As an aside, the crushing strength of a hollow 8- × 16-in (20- × 40-cm) Hadite block is about 110,000 lb (49,900 kg).] Shimming (interior settlement) is often recurrent because the true problem (unstable bearing soil) has not been addressed. However, reshimming is relatively inexpensive, and the rate of settlement decreases with time because, at least in part, the bearing soil beneath the pier is being continually compacted.

2.4 REPLACING SUBSTANDARD FLOOR SUPPORTS

When a structure is supported on wood "piers," or stiff-legs, replacing only a portion of the wood supports with the superior masonry or concrete design is rarely justified. Normally, such a practice would (1) represent little, if any, long-range benefit, (2) represent a waste of money, and (3) possibly prevent a future HUD-insured loan.

Although most foundation repair concerns are directed toward foundations of concrete design, there is the often neglected issue of frame structures on wood foundations. The following discussion addresses several aspects of foundation repair relative to the frame design. The foundation repair may involve (1) leveling of floor systems basically as is, (2) leveling the floors and creating *minimal* clearance between the wood substructure and ground, or (3) providing *adequate crawl space* with concurrent leveling.

2.5 FRAME FOUNDATIONS WITH LIMITED ACCESS

Wood members near to (or in contact with) the ground are obviously influenced more by moisture than those farther removed and protected by air circulation. Contact with moisture encourages the wood framing to warp and rot.

Differential movement over a prolonged period of time exacerbates the same warp. In addition, the limited access to interior floor joists and girders complicates remedial activities. In the end, each condition represents serious deterrents to foundation repair. In other words,

1. Limited or nonexistent crawl space must be addressed. Access is required not only for wood replacement and foundation leveling but also for access to utilities and inspections. In some instances, a crawl space of 18 in (0.45 m) is specified. The latter represents yet another problem, discussed in following paragraphs. For limited access, moderate tunneling can generally provide the access required for simple jobs.

2. Preconditioned warp in the wood substructure must be understood and compromised. Most often leveling is neither practical nor possible.

3. Relative costs for various alternatives to provide access are sometimes prohibitive. Extreme causes could involve removing the floors and in some instances even wall partitions.

Figure 2.2 is a photographic review of a frame foundation before, during, and after repair.

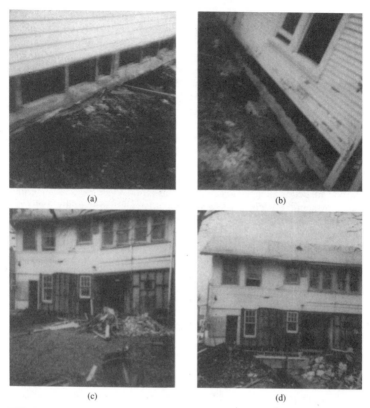

FIGURE 2.2 Repair to frame foundation with limited crawl space. (*a*) Frame foundation on soil is badly deteriorated; (*b*) new wood beam is in place and supported above grade on masonry pads and piers; (*c*) prior to repair—note very extensive sag at doorway and roof eaves; (*d*) after repair—doorway and roof lines are straight.

2.5.1 Providing Access

This situation offers several alternatives, none of which is easy or inexpensive. For example, the interior floors and perhaps some partition walls can be removed as required to provide access to the floor substructure. Concurrently, the skirting can be removed to provide access to the perimeter sill plates. This done, the structure can be leveled (or raised) to create a desired crawl space. Tunneling, although expensive in itself, can be an option. Once the access is provided, masonry pier caps can be installed.

Figure 2.1 shows a typical pier cap design. If the foundation is to be raised to provide the 18-in (0.45-m) crawl space, concern must be given in advance to utility connections to circumvent unnecessary damage to water, sewer, or electrical connections. Hand excavation is expensive. The excavation to provide an 18-in (0.45-m) crawl space for a 1000 ft^2 (93 m^2) house would remove 55 yd^3 (42 m^3). Some contractors figure that a single laborer can tunnel 2.0 yd^3 (1.5 m^3) in an 8-h day. At an hourly rate of \$6.50/h, the example would cost \$5700. Care must be taken in the use of the word *level*. This term is not applicable to foundation repair procedures in general and even less so when old frame substructures exhibit warp.

Alternative choices include the use of steel beams to raise and support the structure independent of the foundations. (The beams are generally aligned to utilize any existing girder/perimeter support system.) As a rule, house movers are better equipped to handle this operation. Once the structure is elevated, two options become available. First, if lateral space permits, the structure can be moved aside to provide access to drill piers and set forms for installation of new concrete piers, pier caps, and perimeter beam. Second, if lateral space is not available, the fact that the structure is elevated will permit some work to be done. Precast masonry piers and pads can be located to support the lightly loaded interior floors. Conventional spread footings, concrete piers with haunches and pier caps, or continuous concrete perimeter beam can be poured in place to support the exterior. Once the foundation has been constructed, the house can be reset.

In summary, for most cases the most favorable option is to remove sections of floor and excavate soil to provide the desired access and minimal clearance between the wood substructure and ground. This approach is often the least expensive. The expensive solutions normally involve those instances where excessive excavation is required for the creation of a complete crawl space, or the structure is literally moved to permit the installation of a foundation. Many problems are resolved once the foundation is raised by whatever means and supported by a masonry or concrete foundation. Access, ventilation, and insulation from the ground are handily resolved. The final goal is to level the floors. This ambition is often compromised because warped wood subjected to no appreciable load is not likely to relax or straighten. A high spot left unsupported might, over time, settle somewhat to project a degree of leveling; however, this is by no means predictable. Generally the pier caps are shimmed to provide a "tight" floor. The bottom line is that the end results are clearly a degree of compromise.

2.6 FOUNDATIONS WITH ADEQUATE ACCESS

This situation provides a much better, less-expensive set of approaches.

2.6.1 All Frame Foundations

If all foundation support members are frame, two options exist: (1) The floor system (including perimeter) can be reshimmed on the existing wood piers, or (2) the wood piers can be completely replaced by masonry pier caps on concrete pads (Fig. 2.1).

2.6.2 Concrete Perimeter with Frame Interior Piers

For foundations with concrete perimeter beams and wood interior piers, the interior wood piers can be either shimmed as is or replaced, as before, with masonry pier caps and pads. Problems with the perimeter beam are classically addressed by conventional underpinning (refer to Chap. 5).

2.6.3 Concrete Foundations

Floor systems with a girder-wood substructure, supported in turn by a concrete pier and beam foundation, represent a fairly easy challenge (Fig. 2.3). The floors are releveled by raising the girders. The girders are in turn held in position by reshimming on the existing pier cap (Fig. 2.1).

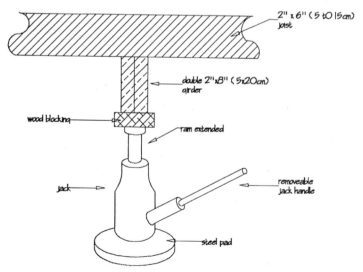

FIGURE 2.3   Jack in place to raise girder.

2.7 EQUIPMENT NECESSARY AND RAISING PROCEDURE

The installation procedures are much the same whether the project is to shim the existing pier caps or to install supplementary ones. The following will assume ready access.

2.7.1 Selection of Jacks

The first equipment to select is the jack needed to provide the force necessary to lift the structural weight. Frequently, Norton or Simplex journal or hydraulic jacks are used for this operation. These jacks are about 13 in (33 cm) in length, have a 5-in (13-cm) extension, and are rated at 25-ton (22,700-kg) capacity. The choice is usually the hydraulic jack, which weighs and costs about one-half of the others. A normal leveling operation could require 4 to 12 jacks.

2.7.2 Positioning the Jacks

The jacks are strategically placed beneath the wood members to be raised (Fig. 2.3). If the soil is soft, the jacks should be placed on a steel plate or wood blocks (or both) to prevent the jacks from sinking into the dirt. A steel plate could be a circular piece, $3/8$ in (9.5 cm) thick, and 20 in (50 cm) in diameter with a handle wielded to one or two sides. The blocks are normally pieces of hard wood 2 to 4 in (5 to 10 cm) thick and of various widths and lengths. The same wood is also used as a block on top of the jack head to both preserve the extension and spread the lifting pressure. Figure 2.4 illustrates jack placement used to install a single girder. Figure 2.5 addresses the problem of installing multiple girders. To raise the floor position, a watch person is stationed inside the area to be raised, and upon his or her command, the four jacks at position B raise the floor about 1 in (2.54 cm). Next, the jacks at A and C are raised about $3/8$ to $1/2$ in (0.9 to 1.25 cm). This process is repeated until the maximum amount of leveling is achieved and then the precast pier caps on pads are installed.

If the desired raise exceeds the extension capacity of the jack, it becomes necessary to (1) temporarily block the girder, (2) compress the ram, (3) add blocks and reposition jack, and (4) repeat original process.

When supplementary floor support is necessary, the process usually involves the addition of girders supported on masonry pier caps, as described above. Figures 2.4 and 2.5 depict such operations. Upon occasion the need to supplement floor joists is also involved. On even rare

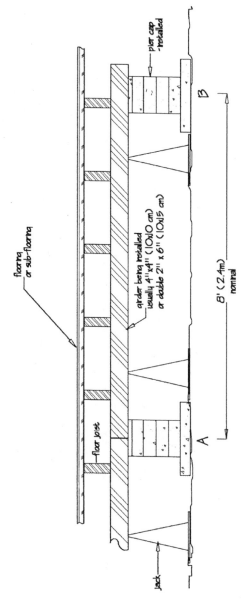

FIGURE 2.4 Installing a wood beam to act as a girder support. Note the splice at pier cap A. Jacks are removed when pier caps have been installed.

2.9

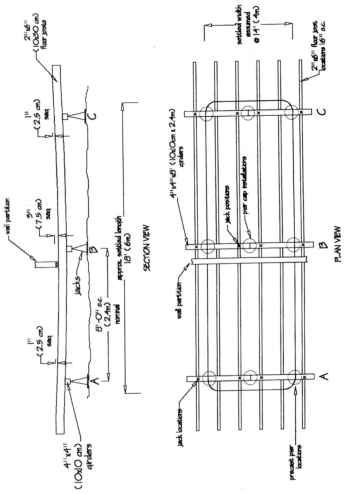

FIGURE 2.5 Positioning supports for leveling floor sag. Refer to Fig. 2.4 for detail at splice.

2.10

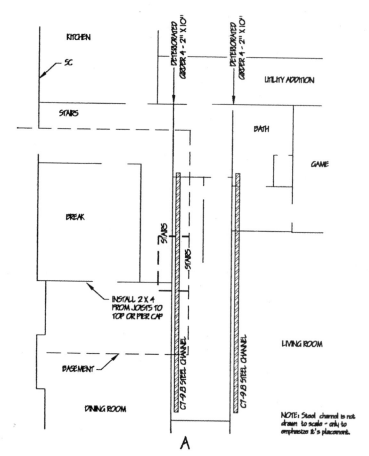

FIGURE 2.6   (*a*) Floor plan showing placement of channel iron.

occasions the responsible approach might require the installation of steel
I beams or channel iron (Fig. 2.6).

2.7.3   Costs

The aforementioned installations are likely to cost the following:

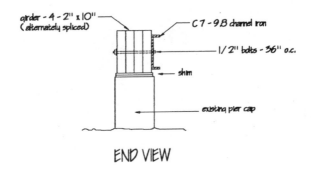

girder - 4 - 2" x 10"
(alternately spliced)

C 7 - 9.8 channel iron

1/2" bolts - 36" o.c.

shim

existing pier cap

END VIEW

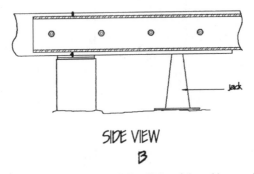

jack

SIDE VIEW
B

FIGURE 2.6 (*Continued*) (*b*) Installation of channel iron to reinforce wood girders.

Installation	Cost, $
Install girders	12/b. ft*
Scab floor joists	12/b. ft
Install masonry precast pier caps; assume maximum height of 18 in (45 cm):	
2- × 2- × 4-in (60- × 60- × 10-cm) pads	90 each
1¹/₂-ft × 1¹/₂-ft × 4-in pads (.45- × .45- × 10-cm)	80 each
The cost for cast-in-place pads or pier caps depends to a large extent on access; assuming at least an 18-in (0.45-m) crawl space:	
30-× 30- × 9-in (75- × 75- × 24-cm) pad	190 each
12-in (30-cm) diameter pier cap	48 each

*b. ft = board feet. One board foot is represented by a 1- × 12-in board 1 ft in length.

The cost for an example installation of channel beams to reinforce joists or girders is much higher than the installation of supplemental wood. This cost is approximated as follows (refer to Fig. 2.6c and Table 2.1):

C-4 (7.25 lb)	$34/linear foot
C-5 (6.7 lb)	$31/linear foot
C-7 (9.8 lb)	$45/linear foot
C-8 (11.5 lb)	$53/linear foot

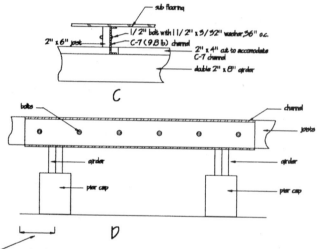

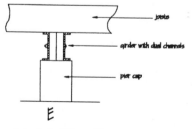

Length past girder / pier cap dependant upon span between pier caps, size, and strength of channel, and structural load. Minimum overlap is generally 12" (3m) .

If wood member sags between support points, it is desirable to jack the joist or girder level prior to drilling the wood for the bolt installation. In severe instances, a supplemental pier cap can be installed or the distance between bolts reduced.

FIGURE 2.6 *(Continued)* *(c, d, & e)* Channel iron used to stiffen joists or girders.

TABLE 2.1 Weights and Dimensions of Structural Steel Channels

Depth of channels, in	Weight/ft·lb	Thickness of web, in	Width of flange, in
	6.00	0.362	1.602
3	5.00	0.264	1.504
	4.10	0.170	1.410
4	7.25	0.325	1.725
	5.40	0.180	1.580
5	9.00	0.330	1.890
	6.70	0.190	1.750
	13.00	0.440	2.160
6	10.50	0.318	2.038
	8.20	0.220	1.920
7	14.75	0.423	2.303
	12.25	0.318	2.198
	11.50	0.220	2.260
8	18.75	0.437	2.260
	13.75	0.303	2.343
	11.5	0.220	2.522
	9.8	0.210	0.209
9	20.00	0.452	2.652
	15.00	0.288	2.488
	13.40	0.230	2.430
	30.00	0.676	3.036
10	25.00	0.529	2.889
	20.00	0.382	2.742
	15.30	0.240	2.600
12	30.00	0.513	3.173
	25.00	0.390	3.050
	20.70	0.280	2.940
15	50.00	0.720	3.720
	40.00	0.524	3.524
	33.90	0.400	3.400

Source: *The Building Estimators Reference Book,* F. R. Walker Co., 1970.

Prices include installation and bolting through the wood joists on 12- to 30-in (0.3- to 0.75-m) centers. Channels are predrilled by the supplier. Prices assume reasonable access and channel total weights are less than about 150 lb (68.2 kg). The unit prices fluctuate slightly depending on spacing and diameter of bolts, access, current steel and labor prices, weight of channel, etc. The cited prices include $1/2$-in (1.25-cm) bolts on 24-in (0.6-m) centers. The material costs for the C-7–9.8 channel iron drilled, cut, and delivered was $6.50 per linear foot. Table 2.2 provides weight and dimension values for I beams. The foregoing was based on the following assumptions:

TABLE 2.2 Weights and Dimensions of Structural Steel I Beams

Sec. no.	Wt/ft, · lb	CSA, in^2	Section width, in	Flange Width, in	Thick, in	Web, thick, in
W-6 x	16	4.74	6.28	4.03	0.405	0.260
	12	3.55	6.03	4.0	0.280	0.230
	9	2.68	5.9	3.94	0.215	0.170
W-8 x	15	4.44	8.11	4.015	0.315	0.245
	13	3.84	7.99	4.00	0.255	0.230
	10	2.96	7.89	3.94	0.205	0.170
W-10 x	19	5.62	10.24	4.02	0.395	0.250
	17	4.99	10.11	4.01	0.330	0.240
	15	4.41	9.99	4.00	0.270	0.230
	12	3.54	9.87	3.96	0.210	0.190
W-12 x	22	6.48	12.31	4.03	0.425	0.260
	19	5.57	12.16	4.005	0.350	0.235
	16	4.71	11.99	3.99	0.265	0.220
	14	4.16	11.91	3.97	0.225	0.200

Source: Armco Steel Corporation, Houston, Texas.

2.15

1. All concrete to be reinforced with #3 rebar (0.9 cm).
2. All prices quoted are based on unskilled labor at $6.50/h and concrete at $55/yd^3.
3. There are no extraneous restrictions or interferences.

Note that the use of structural steel to reinforce or supplement wood substructure members (joists or girders) to support interior floors on residential pier and beam foundations is broadly considered to be a grossly expensive overkill.

2.8 ROTTED SUBSTRUCTURE

As stated earlier, the bigger cost in replacing existing wood floor supports is *access* (see 2.1.C). Once the access problem is addressed, removal and replacement of the wood supports, or stiff-legs, are handled as discussed in Sec. 2.3. The need addressed in this section is often occasioned by rotted decay of original wood. Prudent practice suggests the replacement of *all* supports (perimeter and exterior) once any leveling action is initiated.

2.8.1 Causes of Wood Deterioration*

One of the most common causes of the degradation of building timbers is a decay of the wood structure caused by wood-inhabiting fungi. This infection of the wood frequently alters its physical and chemical characteristics. The extent of the deterioration depends to a large extent on the degree of the decay and the specific effects of the organism producing it. The strength and density of the material will be drastically reduced, and the color will be permanently modified.

During the invasion stage, there is frequently no visible change in wood other than the possible discoloration. Often the condition could go undetected for a period of time. In the late or advanced stages of decay, the wood may become punky, soft and spongy, stringy, pitted, or crumbly, depending on the nature of the attacking fungus and time lapsed. The development and growth of wood-inhabiting fungi requires moisture, air, oxygen, and relatively warm, ambient temperatures. Decay is most prevalent in wood that is either in direct contact with damp ground or located where moisture collects and cannot be readily evaporated. This emphasizes the need for proper ventilation in the crawl space to eliminate moisture. Although individual fungi show a variation

---

*Special appreciation goes to Dr. Don Smith, Biology Department, University of North Texas, for input to this section.

in the precise moisture requirements, it is recognized that a moisture content above the fiber saturation point (25 to 30 percent) is required for optimum development. A moisture content below 15 percent completely inhibits growth (dry). Moreover, if wood in which decay has already started is dried to a moisture content below 20 percent, the development of the fungus will be stopped. However, once this infected wood is again placed in an environment that has a moisture content in excess of 20 percent, decay will generally start again.

Dry rot (or the last stages of brown rot) describes the condition when wood becomes brittle and crumbly. This is usually more obvious as the wood dries. The name *dry rot* is a complete misnomer. No wood will decay while it is dry. The dry rot fungus, which frequently is found growing in comparatively dry locations, is capable of seeking and supplying its own moisture through root-like strands of mycelium. Timbers in touch with moist ground can often provide moisture for decaying areas 15 to 20 ft away. Many engineers suggest wood replacement when the structural timber (sill, joists, girders) shows more than 20 percent damage.

Sometimes mold, mildew, and fungus stains are present on building timber surfaces. Most molds or fungus stains have little effect on the structural integrity of the timber. They are generally surface conditions that are also conducive to those of wood-destroying fungi. However, structural problems related to the invasion of mold and mildew are less preponderant than fungi.

Most wood that has been wet for any considerable length of time probably will contain bacteria. The presence of a sour odor manifests bacteria action. Usually, the greatest effect of cellular dissolving bacteria is that it allows for excessive water absorption, which can promote strength loss in some species of wood. This occurrence is somewhat similar to that described for mold and mildew in that the frequency of serious problems is somewhat limited.

In summary, the decay of timber that causes a structural deficiency is generally termed *wood rot*. These conditions result most often from the development of a wood-destroying fungus that feeds on the wood, affecting the strength and density. This condition normally occurs over a long period of time. Frequently, there are periods when the fungus is dormant and growth and decay are stopped due to ambient temperature and environmental changes. Building conditions that can promote the growth of fungus and the resulting decay of the wood timber include such factors as poor drainage (which results in moist soil), wood in contact with moist soil, long-time plumbing leaks, wood not properly treated or protected, and poor or improper ventilation. Any act that prevents or removes moisture is a preventative measure.

Prudent practice suggests the replacement of *all* supports (stiff-legs; perimeter and exterior) once any leveling action is initiated. HUD encourages this practice.

2.8.2 Costs

The costs to replace a wood member vary from application to application. However, a rough cost estimate for removal and replacements follows:

Joists/girders	$15.00/linear foot ($12 per board foot)
Sill plates	$23.00/linear foot
Pier caps	See sec. 2.4

The cost basis (labor, etc.) would be the same as that specified in the preceding. Refer also to Sec. 2.7.3.

2.9 PRECAUTIONS AND PREVENTION

Chapter 8 presents a detailed discussion of measures generally considered to avoid or minimize foundation failure. The ensuing paragraphs, however, highlight concerns particularly of interest to pier and beam foundations.

2.9.1 Drainage and Ventilation

It is most important to prevent standing water in the crawl space, even if the accumulation is sporadic. Occasionally, construction practices tend to promote this problem (Fig. 1.3). The crawl space is lower than the outside grade. Because water always seeks the lowest level, there exists a natural attraction. Water in the crawl space creates several conditions that are unacceptable. These include (1) an attraction for termites, (2) the buildup of mold and fungus, (3) unpleasant odor, (4) wood rot, and (5) potential foundation damage. Attempts have been made to "mask" certain of these problems by covering the crawl space with polyethylene. This could have some beneficial influence on items 2 through 4 but might actually increase the likelihood of items 1 and 5. The best cure is to (1) establish proper drainage away from the perimeter so water flows away from the structure, and (2) provide sufficient ventilation so any intruding moisture will be promptly removed. The latter might require a forced air blower to increase circulation and/or installation of additional foundation vents. [Building codes suggest 1 ft^2 (0.9 m^2) of vent per 150 ft^2 (13.5 m^2) of floor space.] Flower boxes and curbing can also create sources for water. Flower boxes should have a concrete bottom and drains that direct excess water away from the perimeter beam. Flower bed curbing must also contain provisions to drain water away from the perimeter beam.

2.9.2 Crawl Space

As stated above, the crawl space should be dry and provide access to the foundation, utility lines, ducts, etc. As a rule, the clearance should be at least 18 in (0.45 m). Adequate ventilation should also be a primary concern. Most building codes suggest 1 ft^2 (0.09 m^2) of air vent per 150 ft^2 (13.5 m^2) of floor space.

CHAPTER 3
MUDJACKING (SLAB)

3.1 INTRODUCTION

Slab foundations suffer the problems of both settlement and upheaval. The general approach to the perimeter is much the same as that described in Chap. 5. However, due to design, mudjacking is required to fill voids and raise and stabilize the foundation.[15–17,19,23,36,60,79,86,102]

Routine mudjacking represents an area where the contractor is generally denied any margin of safety. As voids are filled and the foundation raised, pumping must cease when the violating segment of the foundation slab has been restored to desired grade. About the only factor the contractor controls is to fill the voids as completely as possible—to supply as close to 100 percent foundation bearing support as possible. This is discussed in the following paragraphs. Specific site compromises often preclude as thorough filling as might be otherwise desired. For example, it is often decided not to drill holes through ceramic or sheet linoleum floor tile and sometimes the location of utility lines beneath the foundation also interferes with mudjacking. Subject to these restrictions, mudjacking is often not warranted. When a warranty is provided, it is often limited to 12 months. Flatwork (which has no perimeter beam) is often performed with no or limited warranty. Recurrent settlement of foundations properly mudjacked is not too common, less than 1 to 2 percent. Comparatively, flatwork is about 10 times as likely to resettle—often due to erosion of soil beneath the placed grout. Flatwork includes such installations as walks, patio, drives, streets, parking slabs, and pool decking.

For all practical purposes, conventional deformed bar and post-tensioned slab foundations are treated as the same. In the latter case, it sometimes becomes necessary to also repair or retension defective cables.

3.2 EQUIPMENT REQUIRED

Mudjacking is a process whereby a water and soil cement or soil-lime-cement grout is pumped beneath the slab, under pressure, to produce a lifting force that literally floats the slab to the desired position. Figure 3.1 depicts the normal equipment required to mix and pump the grout. The mudjack shown behind the truck is a Koehring Model 50, theoretically capable of pump rates to 10 yd^3/h (8 m^3/h) and pressures to 250 lb/in^2 (1725 kPa). The grout is introduced via small holes drilled through the concrete. The rubber nozzle affixed to the end of the injection hose serves as a packer to contain the grout beneath the slab (Fig. 3.2). The pattern for holes drilled to facilitate grout injection is dictated by the specific job condition and guided by experience. "Down" holes are drilled vertically through the slab surface and used to conditionally raise interior areas of the slab. These holes are located to both avoid unnecessary floor damage and, at the same time, provide the best possible results. Ceramic tile and sheet linoleum floor covering often influence the hole pattern. Other holes are drilled horizontally through the perimeter beam. Where possible these holes are drilled below grade. The back fill then covers them.

Often routine mudjacking is mistakenly referred to as pressure grouting (see Chap. 4). The weight of a typical 4-in (10-cm) -thick concrete slab is on the order of 50 lb/ft^2 (810 kg/m^2). In terms of pressure, this relates to less than 0.35 lb/in^2 (2.4 kPa). Mudjacking the perimeter beam where applicable requires greater pressure. Neglecting breakaway friction, this load could approach 5 lb/in^2 (34 kph). In either event, "high pressure" is hardly as descriptive a term.

Seldom is any uniform grid applicable except in routine raising of open slabs such as pavements, walks, floating slabs (i.e., warehouse floors), or aspects of pressure grouting (see Chap. 4). In fact, often during slab foundation leveling, the hole pattern is adjusted to provide the desired control and results. The floor pattern is controlled by selective use of injection bleed holes. For example, pumping starts at hole A. Holes surrounding the injection site are left open. As pumping continues (to fill voids), the grout appears at one or more of the surrounding holes. To direct the grout to a different location, the bleeding hole(s) is temporarily plugged. Tapered, round wood pegs or rolled-up cement bags can be used as plugs. When voids are filled, all holes except the injection site are plugged. Continued pumping produces the selective raising. Mudjacking gradually proceeds in more or less a circular pattern to the extremities. Upon completion the mudjack holes are patched with a low slump concrete. The patches are raked-off even with the slab surface.

3.3 MATERIALS, COMPOSITION

Grout composition and consistency is an important consideration. A typical grout (28-day compressive strength of 50 to 100 psi, or 345 to 690 kPa) could

(a)

(b)

FIGURE 3.1 (*a*) Mudjacking equipment; (*b*) close-up view of Koehring Model 50.

consist of two sacks of cement [376 lb (170 kg)], 1800 lb (820 kg) of siliceous soil, and 70 gal of water per cubic yard (330 L/m³) of dry mix. The two-sack mix is the general choice for mudjacking. Unless the grout is watered down, this amount of cement provides adequate strength, limits shrinkage, and facilitates flow. The set grout is also friendly to excavation or reentry. In specific cases, richer grouts can be used. For example, in cases requiring increased strength, a four-sack grout might be used [compressive

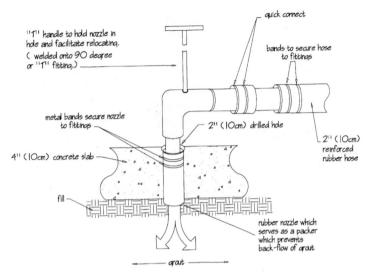

"T" handle to hold nozzle in hole and facilitate relocating. (welded onto 90 degree or "T" fitting.)

quick connect

bands to secure hose to fittings

metal bands secure nozzle to fittings

2" (10cm) drilled hole

2" (10cm) reinforced rubber hose

4" (10cm) concrete slab

fill

rubber nozzle which serves as a packer which prevents back-flow of grout

grout

FIGURE 3.2 Nozzle pack-off.

strength approximately equal to about 400 psi (2760 kPa) at 28 days]. In cases of dam grouting the grout used might be nothing more than cement and water (5 gal/sack), often referred to as a neat cement grout. (Compressive strength approximates 8295 psi after 28 days.)

Consistency is normally varied by adjusting the solids-to-water ratio. A thinner grout will migrate over a larger area, sometimes allowing a greater lifting force at the same pump pressure ($F = Pa$). A thin grout is more prone to bleed out from the work area. A thicker grout (less flow) will be restricted to a smaller area. A substantial lifting force may be possible as a result of increased, confined pressure. In some instances, the thicker grout still tends to escape from the work area. Consider, for example, that situation where the grout flows beyond the foundation or intended work area. This could represent a project involving an interior fireplace (concentrated load) surrounded by a normal slab (low weight and strength), a shallow (or absence) perimeter beam, or the like. If the thickened grout does not provide the solution, other options, in order of increasing difficulty (and expense), are

1. *Stage pumping.* This practice involves the placement of a volume of grout followed by a period of shutdown sufficient for the in-place grout to thicken or perhaps reach initial set. The process is then repeated. Care must be given to limit the shutdown time to prevent the grout from setting up in the hose. Refer also to Sec. 3.4.

2. *Shoring for containment.* Sheet piling (plywood or suitable material) is driven or buried into the soil at the slab perimeter and shored by suitable bracing. Under most conditions the seal between the sheet piling and concrete perimeter is sufficient to contain the grout.

3. *Underpinning.* Underpinning can be used to raise the slab perimeter, followed by mudjacking to fill voids. The underpinning supports the structural load, removing that resistance to grout flow. On occasion sheet piling might again be utilized to further contain the grout.

3.3.1 Variations in Grout Mix

More complicated grout mixes are sometimes specified for specialty mudjacking of slabs such as (1) highways, (2) runways, (3) parking lots, or (4) upon rare occasion, commercial or even residential foundations. These mixtures often involve constituents such as fly ash (pozzolan), cement, sand, silicious soil, surfactants (reduce surface tension or water requirements), lime, bentonite (montmorillonite) sodium chloride, and water.

Note: Proprietary grouts are also specified but only rarely due largely to cost and handling problems. These include such products as polyurethane (Uretrek), Cemill (microfine cement), polyacriylomides, and sodium silicates. Expense and limited improvement over conventional mixes inhibit any wide acceptance of complex mixes, at least for mudjacking as defined herein. Various grouting operations, however, frequently use the special mixes and/or products. See references 8, 17, and 87 and Sec. 4.7.

Fly ash or pozzolan has been utilized as a companion to cement for centuries. [In fact the early aqueducts constructed by the Romans (and existing in part to this day) were made from a cementitious material consisting of pozzolanic earth, $Ca(SO_4)$ and water.] Generally the fly ash is intended to reduce (1) the cost of the cement grout with minimal loss in strength, (2) the grout unit weight, and (3) in some cases, grout shrinkage. The principal disadvantages of the use of this product in mudjacking relate to its very abrasive effect on pump and mixing equipment and the practically nonexistent cost savings. Fly ash-cement grouts can, in fact, be much more expensive than conventional grout, depending upon the volume of siliceous soil replaced by fly ash.

Surfactants generally do not serve any real purpose in mudjacking. The grout is placed essentially through voids (no permeability problem), and consistency (flocculation) can be adequately controlled by the discriminate use of water. Lime is sometimes useful to modify the properties of the grout. However, the contact of lime and sulfates in clay soils can be quite deleterious. See references 17, 48, 87, and 85.

Bentonite has limited use in mudjacking. Principally, this is either as an additive to make harsh solids pumpable (i.e., fly ash or sand) or to reduce friction (see Sec. 4.7.2). Bentonite can also be preswelled to create volume and, upon exposure to cement, will still develop a "set." Saline water (sodium chloride or potassium chloride) will reduce the amount of bentonite swell. This reduces the volume increase and tends to somewhat enhance strength. However, if final strength of the grout product is a concern, special care should be given to both the application and composition of bentonite grouts.

3.4 GROUTING PRESSURE

Grout placement pressure is another concern. Most often this aspect is overrated because foundation slabs represent minimal loads. For example, the weight of a 4-in (10-cm) -thick slab is about 50 lb/ft^2, or 0.35 lb/in^2 (244 kg/m^2). In most residential construction, the added live load plus dead load on the interior slab is minimal, often assumed to be 60 lb/ft^2 (292 kg/m^2). [The perimeter loads are substantially greater, perhaps 600 to 800 lb per linear foot (890 to 1190 kg/m).] However, the perimeters are seldom intended to be raised solely by mudjacking. (This was not always the rule, but expediency and lack of gifted operators has prompted the compromise.) Multiple-story construction increases the weight, but still the loads are relatively low. Hence, most mudjacking is accomplished at very low pressure. Significant pump pressure surges are generally the result of various forms of friction increase—plugged lines, improper grout mixture, or upon occasion, that instance where the member to be raised suffers some form of mechanical binding or resistance. Once the grout leaves the pump, there are few instances that could account for a measurable pressure drop. The most frequent cause is a parted hose or blowout. This is best comprehended by understanding the minimal resistance offered by the slab during mudjacking procedures.

Be very alert to any pumping ceasage. During *routine* mudjacking these stoppages are only long enough to permit moving the nozzle from hole to hole. The duration is short and because this process more frequently occurs during the void-filling phase, the problems are minimal. The pause in grout movement through the hose is subject to several factors that complicate, or in worse-case scenarios, completely prevent continuation of pumping. The factors that contribute to this are

1. The dilatant nature of the grout. With dilatant fluids the apparent viscosity is proportional to the applied stress. When in motion, the apparent viscosity is relatively low. When motion stops, viscosity increases (due to stress) and the grout behaves somewhat as a solid. In this state a disproportionately large force is required to reinitiate movement. The lower the water-to-solids ratio, the more pronounced this problem becomes.

2. Cohesion tends to return to the clay, again increasing apparent viscosity.

3. Some cementatious reactions develop between the siliceous sand and cement. Temperature and time are allies for this reaction (along with the amount of cement and water in the grout).

In possibly difficult situations, it is advisable to clear the pump hose prior to any shutdown. This can be done by purging the grout in the lines. The options are wasting the grout or circulating the grout back into the mixer. Water is most often used to clear the lines. If situations demand it, the lines can be cleared by placing a sponge in the line (at the discharge of the pump) and pumping the sponge through the hose using water. *Do not get in front of the hose*; the sponge might eject at a high velocity. When air is used to displace the sponge, even greater care must be exercised.

3.5 GROUT VOLUME

More significant than placement pressure is the placed volume because this aspect has a direct bearing on project cost. In normal slab-raising operations, an operator who is capable of mixing and pumping on the fly can place up to about 16 yd^3 (12.8 m^3) per 8-h day. For most mudjacking (and grouting) operations, the grout volumes are "wet" yd^3, which correlate to the calculated void volume. Wet volume is computed by including the water. For example, the typical grout mix would contain 1 yd^3 of soil (0.76 m^3), 2 ft^3 (0.57 m^3) of cement, and 9.6 ft^3 (72 gal) of water. This would provide a yield of 1.4 yd^3 of grout per yd^3 of soil. The complexities imposed by foundation leveling reduce this capacity to a maximum of about 10 yd^3 (8 m^3), which corresponds to a volume of about 800 ft^2 (74 m^2) in area and 4 in (10 cm) thick. Void filling (amounting to open-ended pumping) can sometimes reach a placement volume of 20 yd^3 (16 m^3) or so per 8-h day, if material supply and handling can support the quantity. If placement volumes are required in excess of these, the best solution is multiple pumps. For other grouting operations, greater capacities are possible. The material can sometimes be batched and fed to alternative pumps, capable of pumping 25 to 30 yd^3 (19 to 23 m^3) per hour, or 200 to 240 yd^3 (160 to 190 m^3) per 8-h day. Refer again to Chap. 4. Omitting slowdowns imposed by changing injection sites, allowance for set time, materials supply, etc., the placement limitation (imposed solely by pressure) can be explained by the relationship

$$HHP = kBHP = \frac{Q \times P}{7.2}$$

where HHP = hydraulic horsepower
   k = an efficiency constant (frequently 0.75 or less)
   BHP = brake horsepower
   Q = rate of flow, ft^3/min
   P = pressure, lb/in^2

At maximum conditions, this equation shows that as the placement pressure increases, the volume pumped decreases proportionately. Note that the placement pressure is the resistance at the pump cylinder head and not the pressure resistance offered by the member being jacked. Friction pressures developed during normal mudjacking operations far exceed the load resistance required in slab raising. In fact, something in excess of about 150 to 200 ft (45 to 60 m) of 2-in (5-cm) inner diameter pump hose and a thick grout will produce a friction pressure approaching the capacity of the average Koehring machine (see Table 4.4). Figure 3.3 illustrates the mudjacking process.

Note: Some opinions tend to minimize the effectiveness of mudjacking. For example, the *Foundation Engineering Handbook,*[105] states "This method [mudjacking] is rarely successful unless the entire consolidation of the ground has taken place." Such a statement is, at best, misleading, particularly as far as residential or light commercial foundations are concerned. The mudjacking is more "permanent" than the natural supporting soil. Recurrent problems would be the result of continued loss of soil-bearing capacity and not the failure of the mudjacking. (This statement was deleted from the 1991 edition.)

3.6 MUDJACKING TO LEVEL A SLAB FOUNDATION

In leveling a slab foundation the primary concerns are (1) avoiding unnecessary damage to floor covering (generally every attempt is made to limit or eliminate the need for drilling holes through ceramic tile or sheet linoleum), (2) raising the areas that are below the desired grade, (3) ascertaining that all voids are properly grouted, and (4) rendering the foundation to as near as-built condition as practical without creating undue, additional damage to the structure. This is accomplished by selective injection through previously drilled holes. The pumping selection is similar to the procedure established in Fig. 2.5. Lowermost areas are addressed first. Refer also to Sec. 3.2.

3.6.1 Settlement

Slab settlement restoration is a relatively straightforward problem. Here the lower sections are merely raised to meet the original grade, thus completely and truly restoring the foundation. The raising is accomplished by the mudjack method as described above. In some instances where concentrated loads are located on an outside beam, the mudjacking may be augmented by mechanical jacking (installation of spread footings or other underpinning). In no instance should an attempt be made to level a slab foundation by mechanical means alone. Instead of providing interior support (and stabilizing the

(a)

(b)

FIGURE 3.3 Mudjacking a slab. (*a*) Injection hole drilled through perimeter beam; (*b*) nozzle in place and mudjacking in progress.

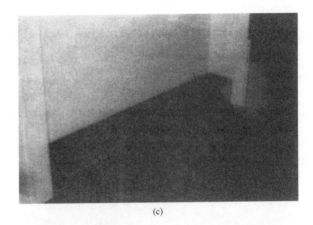

(c)

(d)

FIGURE 3.3 (*Continued*) (*c*) Interior pumping; (*d*) exterior pumping of patio slab.

subsoil), mechanical raising creates voids that, if neglected, may ultimately cause more problems than originally existed. As a rule, residential foundation slabs are not designed to be structural bridging members and should not exist unsupported. (Mechanical techniques normally make no contribution toward correcting interior slab settlement.) Raising the slab beam mechanically and back-filling with grout represents a certain improvement over mechanical methods alone but still leaves much to be desired. This approach loses the benefits of "pressure" injection normally associated with the true mudjack method. In other words, the voids are not adequately filled, which prompts resettlement.

Figure 3.4 illustrates the effect of leveling. In instances involving interior slabs, leveling was accomplished by mudjacking. Perimeters were generally leveled by underpinning techniques. In the "before" pictures, the separations under the wall partition, in the brick mortar, and between brick and door frame and the settlement of the floor slab are obvious indications of foundation distress. In the "after" pictures, the separations are completely closed, illustrating that the movement has been reversed or corrected. In this particular example, the leveling operation produced nearly perfect restoration. This is not always the case, as discussed in various sections in this book. The remaining photos depict the mudjacking process in action.

3.6.2 Filling Voids

Mudjacking is used in a number of other applications either as a remedial or a preventative tool. Chapter 4 covers some of the more spectacular of these. However, there is always the more mundane. This could involve such projects as filling voids to prevent problems. Examples of this could include such projects as filling voids created by water erosion of fill (1) beneath foundations, (2) around utility lines, (3) outside tunnels, or (4) beneath pavement. Another application is back-filling excavations created to provide access to underground utilities.

3.6.3 Sewer Lines

The issue frequently surfaces regarding the interaction between mudjacking and sewer lines. These questions generally boil down to, Did the mudjacking cause or contribute to the sewer drainage? Another issue is whether underpinning contributed to or caused the plumbing damage.

Among insurance companies, a defense has been to claim that the foundation repair companies caused the sewer leaks that then affected the foundation. This position is weak from the standpoint of pure logic. First, the repair contractor would not be on the site unless a foundation problem already

(a)

(b)

(c)

(d)

FIGURE 3.4 Foundation leveling. Before: (*a*) The floor is separated from the wall partition by about 4 in (10 cm); (*b*) the separation in brick mortar is in excess of 2 in (5 cm). After: (*c, d*) In both instances the separations are completely closed. The end results of foundation leveling are not always so impressive.

existed. Second, nothing moves unless forced to do so. With residential foundations on expansive soils, this force is most often water, either natural or (more frequently) domestic. If the basic distress is upheaval, it becomes apparent that the water already existed—long before the foundation repair contractor was called to the scene. The foundation repair contractor could hardly be accountable for causing the leak. Sometimes during foundation leveling, plumbing damage might occur particularly during underpinning. Therefore most foundation repair contractors include a disclaimer that specifically excludes responsibility for damage to underground utilities. The frequency of this occurring is less than 1 percent. Those instances involving perimeter raises over 4 to 6 in (10 to 15 cm) also sometimes cause plumbing damage.

Another variation introduces the allegation that mudjacking causes the sewer damage. If this happens, (1) the line separation would be in vertical (and not lateral) lines and (2) the line would fill with grout during mudjacking. As a precaution a competent contractor runs water during the time that mudjacking is ongoing in areas with plumbing. The water is intended to elutriate the cement from the grout, thus preventing any cementious action. If the sewer does plug, it is quite simple to roto-root the line. This evasive action should be taken quickly—certainly before 24 h. The grout becomes plastic/solid within about 4 h. Beyond this time the grout does not move or flow. In 24 h the grout could attain a compressive strength on the order of 4000 psf (19,000 kg/m^2). If grout does enter the sewer line, it would obviously occur during pumping and not hours, days, or weeks later.

It helps to understand the infrequency of mudjacking creating plumbing concerns when the condition of the repair is understood. First, in the event of upheaval, the slab high points are generally in the vicinity of areas with plumbing. Obviously these locations do not then require additional raise. Second, the plumbing lines, at the location of the fixtures, represent the highest grade elevations within the sewer system. This, plus the fact that the lines are not covered by a large amount of backfill, allows more flexibility in the lines at these points. Nonetheless, most foundation repair companies recommend a thorough utilities test before and/or after repairs.

3.7 RELATIVE COST

As foundation repairs go, mudjacking is one of the least expensive. A 1200-ft^2 (112-m^2) slab foundation with routine settlement [3 in (7.5 cm) maximum] can normally be mudjacked in a day. Based on 1998 U.S. dollars and conditions equivalent to those established in Sec. 2.3, this is something in the range of $1300 to $1800.

Mudjacking is not intended to be a cure-all; it is intended to provide specific benefits. In truth, the rap against mudjacking per se is ill-founded and based on lack of understanding and/or misapplication. As a rule, mudjacking is essential to the proper repair of conventional residential slab foundations. Foundations can be designed to cantilever or bridge void areas. However, such design practices are not common in residential and light commercial foundations. To further complicate the situation, the foundation repair engineer or contractor usually does not know the actual foundation design.

CHAPTER 4
DEEP GROUTING

4.1 INTRODUCTION

Remedial measures to correct foundation failures can, in some instances, require more than strictly underpinning and shoring. The problems addressed in this book will be limited to a depth of about 30 ft. More specifically, failure of supportive elements to the foundation related to loose fill on poorly compacted subgrade materials can be corrected by densification or solidification through a process referred to as deep grouting or, more specifically, compaction grouting. In discussions covered by this book the term *deep* could be misleading. Actually, *deep grouting* is used to delineate grouting from mudjacking. This densification generally creates suitable and competent bearing strata and subsequently stabilizes the structure from future failures. This process is particularly suitable to stabilize land or sanitary fills.

Deep (more correctly, intermediate) grouting involves the injection of a soil/cement/water/grout into a loose or fissured soil subgrade (or fill). This grout is pressure injected to the extent that water and/or air is displaced, voids are filled, and the less-dense material becomes encapsulated. The grout infiltrates and encapsulates the soil and upon curing serves to "solidify" the subgrade basically through cementation and densification.

4.2 INJECTION SITE—LOCATION AND DESIGN FOR PROCEDURE

Injection sites and depths of placement are dictated by soil borings or other means to determine depths and extent of fill as well as load-carrying requirement of the structure. Grout injection sites are then configured in a pattern to provide proper coverage of area and competent placement of materials.

Injections are performed in a series of "lifts" so that all applicable depths of the area receive the necessary quantities of grout. As a rule, the grouting proceeds from the bottom upward. Pumping is continued at each level until either of several predetermined events occur, such as (1) placement of a specific volume of grout, (2) pumping until some pressure is reached, or (3) communication of grout. In the event of communication (or "bleeding") problems, it often becomes necessary to stop pumping and move to another injection site until the grout in place reaches at least initial set. When the latter occurs, pumping can frequently be restarted and carried to a satisfactory conclusion. In some severe instances, the use of a downhole packer becomes necessary to prevent the grout from communicating to the surface. Figure 4.1 provides an example. This setup is particularly applicable when injection shafts are predrilled. Although there are many applications and varying techniques, Sec. 4.7 discusses two case histories of deep grouting procedures.

Fishing is seldom required in this type of grouting because the depths are relatively shallow [generally less than about 30 ft (9 m)] and are seldom in a critical location. This allows lost pipe to be merely abandoned after whatever pipe and fittings can be conveniently recovered. A new injection site is created and the process goes on. When this is not the case, pipe lost in the hole can be fished or recovered by using such tools as overshots, spears, or reverse spirals (to recover augers). This equipment is common to oil-well drilling contractors and contractors engaged in projects such as dam, tunnel, or very deep (100 ft, or 30 m) grouting.

4.3 GROUT COMPOSITION

Specific grouts can involve such products as polyurethanes, polyacrylamide, sodium silicates, cement, cement with admixes (i.e., fly ash or bentonite), etc. The specific product is selected based on costs, problems to be remedied, and placement conditions. In the cases focused on by this book, only compaction grouting (or variations thereof) and the use of cement-type grout will be considered. A broader discussion is provided in many other references such as *Practical Foundation Engineering Handbook.*[17] Most of the "cement" grouts are simply a mixture of siliceous soil-cement and water. Strength and consistency of the grout is controlled by varying the solids-water and cement content. Frequently, this grout will be thicker (less water) and/or contain an increased cement content (higher strength) than conventional mudjack grout.

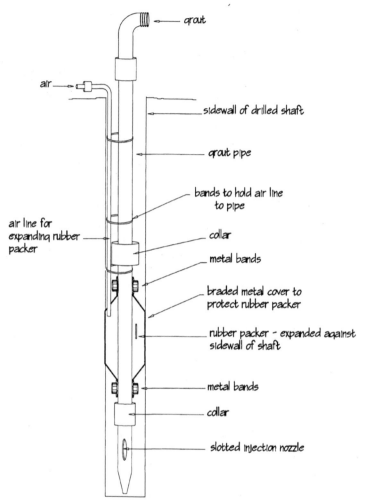

FIGURE 4.1   Downhole packer for grouting.

4.4   MIXING AND PUMPING EQUIPMENT

There is a wide selection of grout mixers and pumps, the choice of which largely depends on the specific project requirements. The operations considered herein are most often handled by the conventional mudjack equipment. Figure 3.1 depicts an example of a common pumping and mixing unit. However, smaller, trailer-mounted concrete or mortar pumps (such as the Whitman) or Colcrete or Moyno pumps can also be quite handy. The latter equipment requires a separate mixer, which can be redimix trucks or conventional concrete (mortar) mixers. Refer to Figure 4.2 for examples. Figure 4.2a depicts a Moyno pump with auxiliary mixers. Figure 4.2b shows a Whitman concrete pump, which relies on a redimix truck to supply the material. The equipment for a particular job is selected based on placement volume and pressure as well as size and access to the work site. Tables 4.1 and 4.2 provides conversion of pressure units. Intermediate grouting as defined herein seldom requires actual placement pressures in excess of perhaps 50 to 100 psi (345 to 689 kPa). However, line friction must also be considered. The latter is influenced by the desired placement rate and size of conduit. Table 4.3 presents data for estimating the pump power required for specific conditions of volume and pressure. The hydraulic horsepower divided by the pump efficiency (often 0.75) gives the corresponding brake horsepower. For example, if Table 4.3 suggests the need for 70.2 HHP (30 yd³/day at 250 psi), the brake horsepower required would be 94 (assuming 75 percent efficiency). If the efficiency factor is 60 percent, the brake horsepower required to deliver the desired HHP becomes 117. Table 4.4 offers *representative* values of friction pressure. These numbers can deviate from those field recorded by a factor of 2 or more dependent largely on the specific grout composition. A more appropriate friction loss value can best be determined experimentally under actual field conditions by establishing the pump rate through an open-ended hose and recording the pressure at the pump discharge.

4.5 PLACEMENT TECHNIQUES

The foregoing and following paragraphs cover some of the questions concerning preplanning for grout placement. As noted, the grout injection pattern varies depending on the particular problem to be resolved. About the only "uniform" criterion seems to be that to produce a grout curtain or consolidated mass requires a staggered injection grid consisting of at least three rows with perhaps a 4-ft (1.2-m) OC (on centers) spacing. (Figure 4.3a shows an injection pattern, although this was created to double as a pattern for subsequent mudjacking.)

4.6 ESTIMATING GROUT VOLUME REQUIRED

Estimating the required grout volume is difficult to do because of the many unknowns. Consequently, most jobs are bid either on a per cubic foot or time and material basis. If some form of "guesstimate" is required, a couple of pointers might be useful:

(a)

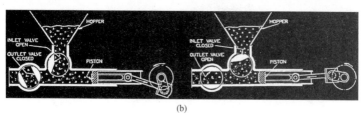

(b)

FIGURE 4.2 Pump and mixing equipment. (*a: Monyo® is a registered trademark of Robbins & Meyer, Inc.*)

1. *Noncohesive soils.* The *theoretical* void in a poorly graded granular material is 40 percent. Giving thought to reality, a workable estimate might be 20 to 30 percent. If the volume to be considered measures 100 ft (30 m) by 30 ft (9 m) by 20 ft (6 m) deep, the soil volume would be estimated at 60,000 ft³ × 0.25, or 15,000 ft³ (424 m³) or 556 yd³. Any void created by erosion would be added to this.

2. *Cohesive soils.* These contain little, if any, porosity under natural conditions. Voids occur as a result of organic decay, consolidation (compaction), or erosion. It is best to attempt to use all information available to "approximate" the void. The grout necessary to handle the problem will equal the volume of the void plus whatever compaction might be desired.

TABLE 4.1 Pressure Conversions—Pounds per Square Foot to Kilopascals

Pounds per square foot	psi	Kilopascals
1	144	0.0479
2	288	0.0958
3	432	0.1437
4	576	0.1916
5	720	0.2395
6	864	0.2874
7	1,008	0.3353
8	1,152	0.3832
9	1,296	0.4311
10	1,440	0.4788
25	3,600	1.1971
50	7,200	2.394
75	9,800	3.5911
100	14,400	4.7880

TABLE 4.2 Pressure Conversions—Pounds per Square Inch to Kilopascals

Pounds per square in	psf	Water head, ft	Kilopascals
1	0.0069	2.31	6.895
2	0.0138	4.62	13.790
3	0.0207	6.93	20.685
4	0.0276	9.24	27.580
5	0.0345	11.54	34.475
6	0.0417	13.85	41.370
7	0.0486	16.16	48.265
8	0.0552	18.47	55.160
9	0.0625	20.78	62.055
10	0.069	23.09	68.950
25	0.174	57.72	172.375
50	0.345	115.45	344.750
75	0.52	173.17	517.125
100	0.69	230.9	689.500

4.6.1 Pricing Grout

The price for grouting is computed based on a per cubic foot in place basis, which varies widely with the specific job condition. A *very* rough number might be on the order of \$8.50/ft^3 (1 ft^3 = 0.02832 m^3) of wet grout.

TABLE 4.3 Hydraulic Horsepower Tables*

Cubic yard per day	Cubic feet per minute	50 psi	100 psi	150 psi	200 psi	250 psi	500 psi
10	0.56	3.89	7.78	11.67	15.56	23.34	38.89
20	1.125	7.8	15.56	23.34	31.12	46.68	77.78
30	1.68	11.67	23.34	35.01	46.68	70.2	116.7
40	2.25	15.6	31.12	46.68	62.24	93.36	155.6
50	2.8	19.45	38.9	58.35	77.8	116.7	194.45
100	5.6	38.9	77.8	116.7	156	233	389
200	11.25	67.8	155.6	233.4	311	467	778
300	16.8	116.7	233.4	350	467	702	1167
500	28	194.5	389	584	778	1167	1945
1000	56	389	778	1167	1560		

*HHP = $Q \times P_d/7.2$, P_D psi and Q, cfm.

P_D = total or delivered pressure = $P + P_c$ where P = friction pressure and P_c = injection or compaction pressure.

TABLE 4.4 Line Friction, ΔP, psi*

Q, cubic feet per minute	Feet of 2′ ID hose					Cubic yards per day†
	50	100	150	200	300	
0.56				0.4	0.6	10
1.125			0.5	0.6	1.0	20
1.68			0.6	0.75	1.2	30
2.25			0.65	0.85	1.30	40
2.8		0.5	0.75	1.0	1.5	50
5.6	0.5	1.0	1.5	2.0	3.0	100
11.23	1.0	2.0	3.0	4.0	6.0	
16.8	1.5	3.0	4.5	6.0	9.0	200
28	3.125	6.25	10	12.5	20	500
56	6.25	12.5	19	25	38	1000

*This table assumes a soil cement grout containing 50 gal H_2O per cubic yard. It provides only a *broad* estimate because the specific viscosity and density of the grout as well as appropriate friction factors are unknown. Where possible the actual friction values should be determined experimentally. The desired delivered pressure is added to the line friction to arrive at the pump head pressure. This total becomes P_D (Table 4.3). (1 ft^3 = 7.48 gal.)

†Day = 8 h.

4.7 CASE HISTORIES

Two jobs were selected to provide a fair overview of intermediate grouting. One project was performed in Dallas, Texas, and the other in Louisiana. Similar problems could be encountered anywhere.

4.7.1 Dallas Hospital Streets

On January 28, 1986, a project was initiated by a Dallas hospital to grout an abandoned waste line that had experienced a washout to a depth of approximately 12 ft (3.6 m) below street level. The washout was created as a result of a 4-in (10-cm) water main break and had undermined the bearing soil of the concrete drive above. The process selected was to fill in the void, restoring the bearing support to the road followed by mudjacking the pavement.

Deep grouting involved the drilling and placement of 2-in (5-cm) outer diameter grout pipes to a predetermined depth of approximately 9 ft below finished grade of the road for placement of grout. Conventional mudjacking equipment was used to mix and pump the grout.

To assure competent placement, holes were drilled in the street surface along the sewer line at 6- to 8-ft (1.8- to 2.4-m) centers (see Fig. 4.3b). The grout pipes were driven to the proper depth by a pneumatic pavement breaker. Connections were fitted to the pipes following placement. A reinforced rubber hose was connected and then attached to a Model 50 Koerhing mudjack. A soil cement grout was pumped through the grout pipes into the subgrade. Grout placement was performed along the sewer lines area to refusal, which assured adequate filling of voids and restoration of bearing support. Upon refusal, it was possible to raise the settled areas of the road and parking garage ramp approximately 4 in (10 cm).

The area treated encompassed 10,000 ft^2 (930 m^2) and required the placement of approximately 38 yd^3 (29 m^3) of grout. Work of this magnitude required 6 days (four-person crew) and was successfully completed without incidence. Refer to Fig. 4.3a and b for more detail.

4.7.2 Louisiana Salt Mine—Hoisting House

In late 1988, a project was initiated by a salt mining company to level and stabilize the production hoist house at a Louisiana facility. The foundation was constructed with a perimeter beam 1 ft × 4 ft (0.3 to 1.2 m) thick in depth with a 1- × 4-ft wide footing poured integrally at its base (refer to Fig. 4.3b). Twelve-inch (0.3-m) -diameter piers had been placed to a depth of 40 ft (12 m) through the overburden soil to the lower salt dome. The interior floor was a 4-in (10-cm) -thick steel-reinforced concrete "floating" slab.

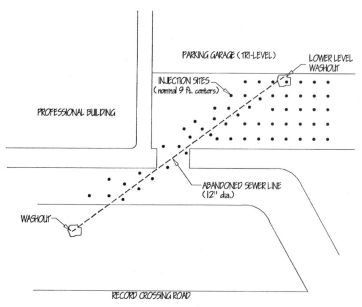

FIGURE 4.3a Dallas Hospital streets, injection pattern.

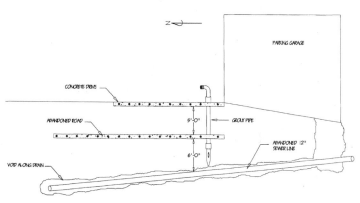

FIGURE 4.3b Hospital—grout pipe placement.

Over the years, groundwater had penetrated the pier shafts, eroding the dome. The erosion of the salt in the pier shafts undermined the support to the production hoist house foundation, creating significant failure to the structure. Principally, the overburden material consisted of quicksand, with some loose gravel. It became apparent that the most practical approach to repair would be to strengthen the overburden soils through deep grouting to create a suitable and competent subgrade. Due to space limitations and the design of the original beam, the proposed repair procedure required installation of spread footings with deep grout pipes cemented in the pour to facilitate raising and sustaining the foundation beam. Attempts to raise the perimeter were unsuccessful due to the enormous structural load (see Fig. 4.4). Consequently, excavation along the entire outside foundation for removal of the 2-ft (0.6-m) overburden on the strip footing (refer to the figure) was performed. To further complicate matters, a beam similar to the one described above was found to transect the width of the interior slab.

Leveling efforts (underpinning and mudjacking) were attempted prior to the deep grouting of the footings to test both the strength of the subgrade and load of the structure. Leveling at this point was not possible. The beam could not be raised without risk of damage.

Grout pipes were placed as necessary throughout the interior slab to depths of approximately 15 ft (4.5 m).

As mechanical raising of the beam was attempted, simultaneously a bentonite slurry was injected on the back side of the beam along with deep grouting to fill voids and densify the subgrade. The bentonite slurry reduced friction, and the combination of these three procedures created conditions that allowed the raising of the beam footings. Grouting was performed to refusal with two "lifts" at 10 and 5 ft (3 and 1.5 m) below slab level.

Once the deep grouting was completed, mudjacking was performed on the interior slab to complete the leveling operations.

The production hoist house foundation area covered approximately 3500 ft^2 (325 m^2) and required installation of 31 footings and removal of over 130 yd^3 of soil around the perimeter. The deep grouting operation necessitated the placement of 234 yd^3 (99 m^3) of soil/cement grout. The pump house is currently in full operation.

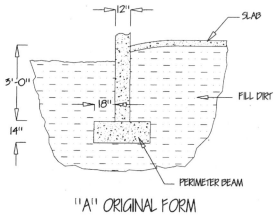

"A" ORIGINAL FORM

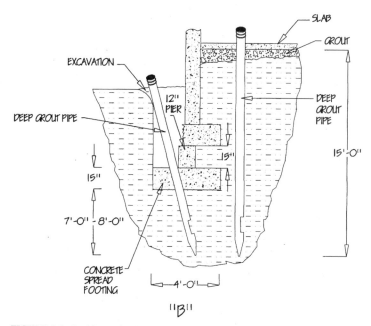

"B"

FIGURE 4.4 Louisiana salt mine.

CHAPTER 5
UNDERPINNING

5.1 INTRODUCTION

A multitude of options have been tried over the years for underpinning foundations. These have included (1) conventional steel-reinforced concrete piers, (2) steel-reinforced concrete spread footings, (3) hydraulically driven steel minipiles, (4) mechanically driven steel minipiles, (5) screw anchors, (6) hydraulically driven concrete cylinders, (7) the ultraslim drilled concrete pier, and (8) the hydropier, which is perhaps the least effective of the lot. Generally speaking, underpinning is relegated to the perimeter beam. On rare occasions, such as settled interior fireplaces, underpinning may be done inside the structure. When this does occur, it is quite expensive and requires breaking out sections of slab floors or removing wood flooring and subflooring in the case of pier and beam foundations. An interior underpin often costs as much as 5 times more than those placed at the perimeter. Tables 5.1 to 5.4, in Sec. 5.6, provide useful data on concrete and rebar.

5.2 CONVENTIONAL DRILLED-SHAFT PIERS

The *conventional drilled-shaft steel-reinforced concrete piers* pose many advantages. The optimum diameter for this pier is 12 in (30 cm) when used with lightly loaded structures (i.e., residential construction).[26,44] The piers are normally spaced on 7- to 9-ft (2.4- to 2.7-m) centers. The shafts normally extend to a depth of (1) adequate bearing capacity, (2) undisturbed native soil, (3) below the local soil active zone (SAZ), (4) rock contact, or (5) acceptable and specific site conditions. Frequently the accepted depth is 9 to 15 ft (2.7 to 4.5 m) below surface. The shafts can be straight or belled. Belling may be required to control upheaval or, in some cases, to enhance the bearing capacity of the pier in substandard soil. The reinforcing steel is at least two #3s but can be as many as four #4s or #5s. The piers are poured monolithically with a haunch usually 30×30×12 in (0.9×0.9×0.3 m), which is also steel reinforced and integrally tied into the pier shaft.[15-17] After adequate concrete curing time, the haunch is used as a base from which to raise the perimeter beam. Once the beam is mechanically jacked to proper grade, a form is set and a pier cap poured. (Although the pier is classically in compression, a rebar is generally centered in the cap mostly as a control for any lateral movement.) The concrete is poured into contact with the foundation beam (see Fig. 5.1).

This serves two important purposes: First, the concrete cap is in intimate contact with the entire irregular beam surface, and second, the use of shim material is avoided. The concrete is not subject to deterioration as shim materials are.

Figure 5.1 presents two options that represent acceptable pier designs. The designs are verified by a simple study of pier moments and haunch bearing. The oversimplified mathematical analysis shows the piers to be safe for the representative conditions. It is always wise to subject any underpinning approach to the same scrutiny. Along these lines refer also to Fig. 5.12. Table

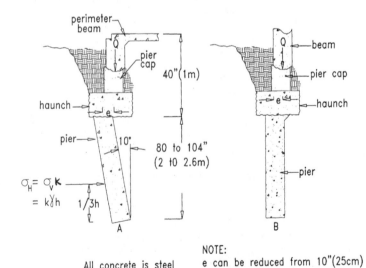

FIGURE 5.1 Two acceptable drilled piers.

5.5, in Sec. 5.6, provides data that might be useful for this purpose. The following analysis does not address the structural competency of the pier itself because this has already been established. The principal concern is to evaluate the soil's capacity to resist the loads.

For study purposes, the structural load was assumed to be 500 lb per 1 ft of beam, the soil unit weight was assumed to be 110 lb/ft³ (17 kN/m³), and the soil's unconfined compression strength was assumed to be 1500 lb/ft² (7320 kg/m²). For *bearing-haunch only,*

$$Q_L = 500 \text{ psf} \times 8 \text{ ft} = 4000 \text{ lb, or 4 k}$$

$$Q_R = 1500 \text{ lb/ft}^2 \times 2.5 \text{ ft} \times 2.5 \text{ ft} = 9375 \text{ lb}$$

$$\text{Safety factor (SF) } 9375/4000 = 2.3$$

where Q_L = wt/ft × ft = load
Q_R = $q \times a$ = resistance

For *moments: piers only,*

A	B
$Q_L = 500 \text{ plf} \times 8 \text{ ft}$	$Q_L = 4 \text{ k}$
$= 4000 \text{ lb} = 4 \text{ k}$	
$M_L = e \times Q_L = 7 \text{ in} \times 4000 \text{ lb}$	$M_L = 10/12 \text{ ft} \times 4 \text{ k} = .833 \text{ ft} \times 4 \text{ k}$
$= 28,000 \text{ in·lb} = 2.3 \text{ ft·k}$	$= 3.3 \text{ ft·k}$

(*e* can be reduced by shaving inside of pier shaft.)

Q_L, M_L results of load;

ϕ = 35 percent

For *soil resistance to lateral movement,*

10-ft pier, 12 in diameter, M_R = moment resistance

$\sigma_H = K_a \cdot \gamma \cdot H$, where K_a = coefficient of passive stress.[65]

$\sigma_H = 10 \text{ ft} \times 110 \text{ lb/ft}^2 \times 0.33 = 363 \text{ lb/ft}^2$; assume $K = 1.0$[16,17]

$P_a = 1/2 \ (363) \ (10) = 1815 \text{ lb/ft}^2$

$M_R = (1) \ (1815) \ 6.7 \text{ ft} = 12,160 \text{ ft·lb}$

$= 12.16 \text{ ft·k}$

The safety factors are

<table>
<tr><td align="center">A</td><td align="center">B</td></tr>
</table>

$$SF_A = \frac{12.16}{2.3} = 5.3 \qquad\qquad SF_B = \frac{12.16}{3.3} = 3.7$$

If the pier depth is increased to 12 ft, the M_R for examples A and B becomes 14.5 and the factors of safety increase to 6.3 and 4.4, respectively.

Obviously, the safety factors are sufficient for the underpinning being considered. The safety factors for the straight shaft can be improved to equal or exceed that shown by the raked pier by drilling the pier further toward the centerline of the beam. In fact, e can be handily reduced to less than 6 in (15 cm).

The presence of the haunch increases the movement resistance of the soil as the presence of the pier increases the load capacity. Once the soil bearing is restored to the foundation beam between pier locations, the load on each pier is *decreased* by a factor of about 8, and the resultant movement on the pier is reduced by the same amount. The raked pier (A) can be drilled with a full auger bit but the straight shaft (B) cannot. This makes the A shaft a little quicker to drill.

5.2.1 Drilling Pier Shafts

Figure 5.2*a* through *c* depicts equipment used to drill piers. The truck-mounted unit represents the quickest and least-expensive means for drilling but requires access and headroom. (The expense advantage becomes even more pronounced as the pier depth and diameter increase.) This equipment can drill rock. The truck-mounted drill is also used to provide shafts for deep grouting, sometimes to a depth of 100 ft (30 m). The tractor rig is effective when access is limited, shaft diameters are 24 in (0.6 m) or less, and depths are less than about 30 ft (9.1 m). This equipment can cut soft rock and bell a 24-in (0.6-m) shaft to 42 in (1.4 m). The latest equipment is the limited-access rig that is used in those cases where access is critically limited.

This equipment can drill with 7-ft (2.1-m) head clearance and no more than 4- to 5-ft (1.2- to 1.5-m) lateral or surrounding access. Drill stems and extensions are generally 4 or 5 ft (1.2 to 1.5 m) in length. This necessitates considerable hand work running or pulling the bit. The rig pictured is capable of drilling a 12-in (0.3-m) shaft to a depth of 20 ft (6 m) and providing a 24-in (0.6-m) bell. The unit shown is capable of delivering 1600 ft·lb (2170 N·m) torque, 5000 lb (2300 kg) lift and crowd, and about 200 maximum rpm. The deterrents are the time required to drill (and bell) a pier and the inherent

(a)
(b)

FIGURE 5.2 Pier drilling equipment. (*a*) Truck-mounted drill; (*b*) tractor drill.

inability to penetrate rock to any degree. The machine can, however, drill smaller-diameter straight shafts to relatively shallow depths at reasonable costs. For example, a 12-in (0.3-cm) -diameter pier can be drilled to 12 ft (3.6 m) for roughly $130 more than that for the truck rig and $90 more than that for the tractor-mounted machine. The cost to drill a limited-access pier is slightly higher than that associated with the excavation for a spread footing. The same pier can be drilled with either of the other machines at a cost below that required to excavate the spread footing.

Unless the truck rig remains on paved surfaces, considerable yard and landscaping damage could result. The tractor rig also tends to create similar (though less) surface disturbances. This is especially true where the yards are less than dry. Figure 5.3 illustrates the drilling sequence for pier-shaft preparation. Figure 5.4 depicts the pier construction, and Fig. 5.5 pictures the haunch and final raising. Drilled shafts must be poured as quickly after drilling as possible. Every attempt possible should be made to pour concrete into shafts on the same day they are drilled.

In some parts of the United States (Florida, for example) the pier is poured by pumping a concrete grout (cement-sand-water) through the drill stem as the bit is pulled. This could be useful if sloughing or water intrusion is to be encountered. Rebar is placed in the shaft after the bit is removed. The

(c1)

(c2)

FIGURE 5.2 (*Continued*) Pier drilling equipment. (*c*1), (*c*2) Limited-access drill.

(d)

FIGURE 5.2 (*Continued*) Pier drilling equipment. (*d*) Lighter limited access equipment is also available. One such model, the Big Beaver, weighs only 500 lb (230 kg), has a lifting capacity of 1650 lb (760 kg), and can produce 398 ft·lb (540 N·m) torque. The company advertises a maximum bit size of 18 in (46 cm) with depth capability to approximately 20 ft. The power unit is located at lower right. This equipment is basically a substantially down-sized copy of the limited-access equipment depicted in (*c*1) and (*c*2). The significant difference is the Kelly drive system. The Big Beaver utilizes a screw drive, and the larger limited access drill uses a chain drive.

concrete mix consists of approximately 1034 lb (469 kg) of cement, 55 gal (208 L) of water, and 20 ft^3 of sand per cubic yard of concrete. These piers, 12-in (3-m) diameter to 12 ft (3.6 m), cost the customer about $1000 (U.S. 1998 dollars).

5.2.2 Designing the Pier

Generally, the piers or pilings depend upon end bearing and skin friction (except in high-clay soils) for their support capacity. In expansive soils, the design contribution for skin friction is disregarded for the top 7 ft (2.1 m) or so of the shaft. Piers or pilings are normally extended through the marginal soils to either rock or other competent bearing strata. This obviously enhances and satisfies the support requirement. The use of piers (or pilings), therefore, is generally restricted to instances where adequate bearing materials can be found at reasonably shallow depths. Generally speaking, for residential repairs, a competent stratum depth below about 20 ft (6.0 m) resists the use of the pier (or pil-

(a)

(b)

(c)

FIGURE 5.3 Drilling sequence used for preparation of concrete piers for underpinning. (a) Drilling the pilot hole for haunch with 24-in bit. (b) Drilling the 12-in-diameter shaft to 15-ft depth. The man at right is enlarging and shaping the haunch. (c) Haunch and pier ready for reinforcing steel and concrete.

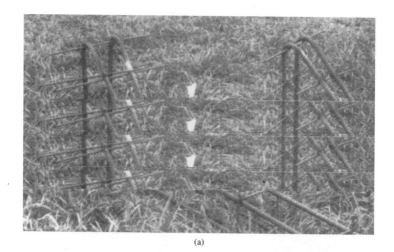

(a)

(b)

FIGURE 5.4   Constructing the concrete pier. (*a*) Rebar caged, ready to place into pier hole (commercial application); (*b*) pouring the concrete.

ing) technique. This depth concern might be lessened, however, by such actions as enlarging the haunch, belling the pier shaft, compaction grouting, or some combination of these. Greenfield and Shen[44] suggest an optimum pier diameter of 12 in (0.3 m). The theory is to minimize pier heave, provide adequate bearing, and, at the same time, facilitate proper steel reinforcing. [It is impractical, if not impossible, to utilize caged rebar in a pier diameter of less than about 10 in (25 cm).] The pier can be belled whenever the soil-bearing capacity is low or questionable. The same reference also notes that pier spacing should be the maximum consistent with beam design, load requirements, and site characteristics. The logic herein is to provide the maximum safe load (surcharge) on the pier to counter expansive soil heave. (Data published by Budge et al. suggest that increasing the surcharge load by a factor of 2 reduces swell potential in pure montmorillonite by a factor of 4.28.[22]) Although Greenfield and Shen principally address original construction, the same design concerns are applicable to remedial procedures.

5.2.3 Raising the Perimeter Beam

Figure 5.5a illustrates a pier cap in place. The beam is raised by the same technique as that described in Chap. 2. The jacks used are generally 25- to 35-ton (22,700- to 31,780-kg) Norton or Simplex journal jacks or hydraulic jacks of similar capacity. Aluminum body 25-ton journal jacks cost slightly less than $1000 and weigh about 40 lb (18 kg). The 25-ton hydraulic jacks cost about one-fourth and weigh about one-half of this. A variation is to use hydraulic rams with an auxiliary power source (such as the unit shown in Fig. 5.5). By use of a manifold and selective control valves, multiple rams can be used simultaneously to literally "float" the structure to the desired grade. (The jacks in other procedures are individually and manually operated.) Occasionally, the need arises for greater jack capacity. When this occurs, the first approach is to work several jack locations simultaneously. [Essentially, the pier locations are spaced about 8 ft (2.4 m) apart.] If this doesn't work, 50-ton jacks (or larger) can be attempted. Special attention must be given to the configuration and load capacity of the concrete beam. It is important not to crack the beam.

Once the beam is satisfactorily raised, a steel pipe section is wedged in the space between the base and the haunch (to the side of the jack). The jack is then removed and a sonotube set in place to act as a form for the concrete pier cap. The pier cap is poured and the excavation back-filled.

5.2.4 Cost of Pier Construction

The material used to prepare the pier system shown in Figure 5.6a would be

(a)

(b)

FIGURE 5.5 Underpinning and raising the foundation. (*a*) Haunch cured and jack in place for raising the perimeter beam. (*b*) The final step—the beam has been raised and the pier cap poured. The steel pipe beside the pier cap is intended only to provide temporary support to the beam until the concrete pier cap cures.

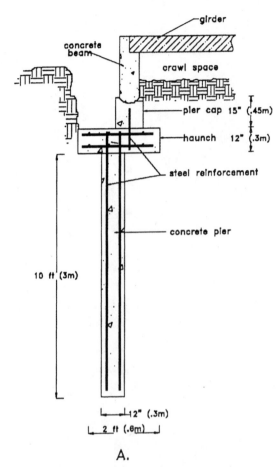

FIGURE 5.6 Raising the perimeter beam.

Concrete: All concrete is 3000 psi

Pier:	0.785 ft² × 10 ft	= 7.85 ft³
Haunch:	2 ft × 2 ft × 1 ft	= 4 ft³
Pier cap:	0.785 ft² × 1.25 ft (15 in)	= 0.98 ft³

12.8 ft³ (0.475 yd³ or 1.19 m³)

Steel: All bars are #3s (3/8 in, or 0.9 cm)

Pier: 4 × 10 ft × 4 × 2 ft (ties) = 48 ft

Haunch: 12 × 1.5 ft   = 18

Pier cap: 1 × 1.5 ft   = <u> 1.5</u>

                                    67.5 ft (20 m)

The cost for the installation of the 12-in (0.3-m) -diameter piers, spaced as shown in Fig. 5.6*b*, could be estimated as follows. Assuming access and no costly delays or interference cause by others, the typical pier should cost about $300 to $500 (U.S. 1998 Dollars). This also assumes that the number exceeds the company minimum (which is often 7 to 10 units). The cost for less than three or four piers escalates by a factor of 1.5 to 2.0. (All prices are based on unskilled labor costs of $6.50 per hour and concrete at $55/yd³.)

5.3 SPREADFOOTINGS

Typically, the spread footings consist of (1) steel-reinforced footings of suffi-cient size to adequately distribute the beam load and poured to a depth rela-tively independent of seasonal soil moisture variations and (2) a steel-reinforced pier cap tied to the footing with steel and poured to the bot-tom of the foundation beam (Fig. 5.7). Design and placement of these spread footings is critical if future beam movement is to be averted. Nominally the footings are placed on 8- to 9-ft (2.4- to 2.7-m) centers. The footing design must consider the possible future problems of both settlement and upheaval and should be of sufficient area to develop adequate bearing by the soil. Often the pad is 3×3×1 ft (0.9×0.9×0.3 m) thick, located at a depth of 30 in (0.75 m) below the perimeter beam. The pier cap should be of sufficient diameter or size to carry the foundation load. It also should be poured to intimate contact with the irregular configuration of the undersurface of the beam. Precast masonry, steel, or wood shims should not be considered as pier cap material. The photographs in Fig. 5.8 depict actual field development of a typical spread footing. Figure 5.8*a* shows a spread footing base pad poured in place. Figure 5.8*b* illustrates the pier poured. The jack used to raise the beam is still in place to the right of the pier, and the steel pipe used to temporarily support the beam until the concrete pier cures is evident to the left of the pier.

The principle of the footing design is to distribute the foundation load over an extended area at a stable depth and thus provide increased support capacity on even substandard bearing soil. The typical design represented by Fig. 5.7 provides a bearing area of 9 ft² (0.8 m²). Effectively, a load of 300 lb/ft² (1465 kg/m²) applied over 1 ft² (0.09 m³) requires a soil-bearing strength of 300 lb/ft²

REMEDIAL SKETCH—PIER LAYOUT
construct 15–12" diameter concrete piers 6'–9' o.c.
10'deep, or to rock
mudjack In affected areas as required to fill volds.

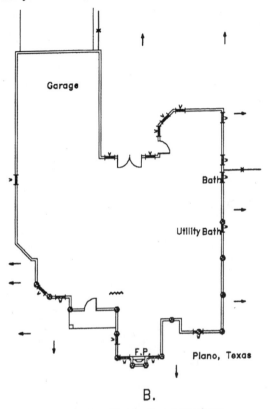

B.

FIGURE 5.6   (*Continued*) Raising the perimeter beam.

(1465 kg/m²). The same load distributed over 9 ft² requires a soil-bearing strength of only 33 lb/ft² (161 kg/m²). Expressed another way, a 9-ft² spread footing on a soil with an unconfined compressive strength (q_u) of 1500 lb/ft² (7320 kg/m²) provides a load resistance of 13,500 lb (6136 kg) ($Q_r = q_u A$). This capacity exceeds the structural loads imposed by residential construction by a wide margin. Generally, the diameter of the pier cap is greater than, or at least equal to, the width of the existing beam. The form for the pier cap must extend outside the beam to permit placement of concrete. Because the pier is essentially in compression, utilizing the principal strength of concrete, its design features are not as critical as those for the footings.

As previously stated, the spreadfooting should be located at a depth sufficient to be relatively independent of soil moisture variations due to climate conditions (SAZ). In London, this depth is reportedly in the range of 3 to 3.5 ft (0.9 to 1.1 m).[54] In the United States, this depth is reportedly in the range of 2 to 3.5 ft (0.6 to 1.1 m).[102] In Australia the depth is considered to be less than about 4 ft (1.25 m).[52,53] In Canada, this depth has been reported to be as shallow as 1 ft. (0.3 m).[97] Along this line of thought, Fua Chen presented a paper that questions the reliability of theoretical approaches to predict heave.[27] The heave prediction methods are based on an assumed depth of wetting, which varies considerably among investigators. Chen also suggests that heave predictions are generally much greater than those actually measured in the field. Does this mean that the seasonal depths of soil moisture change

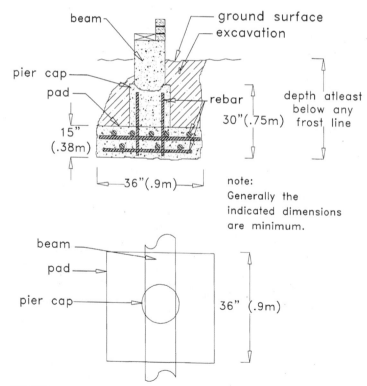

FIGURE 5.7 Spreadfooting.

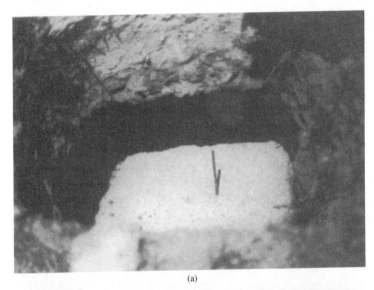

(a)

(b)

FIGURE 5.8 (a) Concrete pad is poured in place (note rebar in place). (b) The concrete pier cap has been poured on the pad to support the foundation beam. The jack used to raise the beam is still in place, and the steel pipe used as a temporary support is in place.

are, in fact, considerably less than values normally assumed? Nonetheless, proper soil moisture maintenance would ensure the stability of the footing with minimal concern for the effective "active depth" or depth of ambient soil moisture variation.

One concern that might arise is the effectiveness of the pad to support the structural load. Generally, this is a nonissue because of the enormous load capacity of the pad (9 ft^2, or 0.81 m^2). However, it can be noted that if the pad cracks under loading, it has failed. If the pad sinks, the bearing soil has failed.

5.3.1 Construction of Spreadfooting

The procedure for construction and implementation is about the same as that outlined in Sec. 5.2 for a drilled pier. First, the excavation is dug as shown in Figs. 5.7 and 5.8. Soil Sta is poured into the excavation as a safeguard against future upheaval should some source provide water to that depth. Soil Sta is discussed in some detail in Chap. 6. Next, the spread footing is poured and allowed to cure. Generally, the footing thickness is in the range of 12 to 15 in (0.3 to 0.38 m). This is much thicker than the structural demand would require. However, from a practical standpoint, the design of the spread footing depends on choosing a depth reasonably free of seasonal moisture variations (30 in, or 0.75 m, for Dallas, Texas, conditions). The 9 ft^2 (0.09 m^2) bearing area allows for a safety factor and permits the same design to be used in areas with soils of lower safe bearing strengths. The base of the spread footing, shown in Fig. 5.7, is about 30 in (0.75 m) below the base of the beam and generally over 40 in (1.0 m) below grade, depending upon the depth of the beam. Next, the clearance required for the jacks is about 15 in (0.38 m). This space permits the use of a wood block on the jack head to help distribute the load and prevent the jack from slipping off the concrete beam. Hence, the thickness of the spreadfooting is determined by subtracting 15 in (0.38 m) from 30 in (0.75 m), which is 15 in (0.38 m). The completion of the installation is as described for the drilled pier.

5.3.2 Cost

The material required for the spreadfooting in Fig. 5.7 would be

Concrete: (3000 psi)

Spreadfooting: $2^1/_2 \times 2^1/_2 \times 1.25$ = 7.8 ft^3

Pier cap: 0.785 ft$^3 \times 1.25$ ft (15 in) = <u>0.98 ft^3</u>

8.78 ft^3 (0.3 yd^3, or 0.8 m^3)

Steel:	(#3s)	
Spreadfooting:	6×2	$= 12$ ft
Pier Cap:	$1 \times 1.5 =$	1.5 ft
		13.5 ft (4 m)

The installed cost for a spreadfooting would be something like $300 to $525, based on the same assumptions as those expressed for the drilled pier.

5.4 ALTERNATIVES

Over the past few years several alternative underpinning methods have been introduced. The following paragraphs present a brief glance at many of these, along with a critical review of their individual strengths and weaknesses.

The practice of underpinning without proper mudjacking has become a matter of litigation. A home owner filed suit against a repair company on the general grounds that the "addition of pilings changed foundation from soil supported concrete slab to slab supported by deep perimeter pilings." (Certainly without proper mudjacking to restore the bearing, the plaintiff would be correct in these allegations.) The Texas Court agreed with the plaintiff.

5.4.1 Steel Minipiles

The *steel minipiles* have been installed eccentric (the majority), concentric, concrete filled, equipped with helix(s), etc.[15–17,21,23,39,57,77,80,86,91] As a rule the minipiles are located on 3- to 6-ft (0.9- to 1.8-m) centers. Most minipiles, regardless of design, have not appeared to enjoy any real success when used in areas with expansive soils.[16,23,39,44,80,86] (Figure 5.9 shows photographs of typical failures in minipiles.) First, the eccentric minipiles are perilously subject to failure due to bending moment and/or lateral stress.[15–17,23] Second, the weight of the structure serves as the reaction block to *hydraulically* drive the pier. Once the resistance on the pier *exactly* equals the weight of the structure, upward movement of the foundation occurs.[16,23] This relates to a 1.0 margin of safety. It is true that piers can be "superloaded" by selective driving; however, this can subject the beam to excessive shear stress and possible ultimate failure. [The safety factor problem can be alleviated by mechanically driving the pipe (impact or torque); however, the other noted deficiencies remain serious concerns.]

Bolting the lift bracket to the perimeter beam often creates many structural concerns. Figure 5.10 shows both severe damage to the perimeter beam and failure in the attachment of lift bracket to perimeter beam. Another problem is the screw anchors that are literally screwed into place. This action disturbs the soil traversed by the screw. The disturbed soil, in effect, acts as a "wick" that tends to pull moisture down to the lowest helix. This has caused serious foundation heave in both new construction and remedial applications.[39] Refer also to A. Ghaly and A. Hanna "Uplift Behavior of Screw Anchors," I and II, *Journal of Geotechnical Engineering,* May 1991.[*] Also, minipiles can be mechanically driven by pneumatic hammering at the driving end inside the pile. A cushion of sand prevents or minimizes damage to the end of the pile. This approach allows for both better alignment and a margin of safety. These piles still suffer the other problems inherent to steel minipiles. And minipiles can be driven concentric to the load. Reference Freeman Piering Systems, St. Louis, Missouri.[16] These afford better alignment but also suffer the same inherent defects common to steel minipiles. These introduce two other serious concerns (i.e., the persistent requirement for shoring and an inflated cost due to the additional excavation; see Fig. 5.11). These piers often cost in excess of $1000 each.

5.4.1.1 *Analysis of Eccentric Pile Driving.*

The mechanics of pile driving is an issue of great concern, particularly in light of behavior patterns disclosed in Figs. 5.9 and 5.10. Figure 5.12 depicts a simplified analysis of pipe behavior during eccentric driving.

The first pile section would be forced outward (away from load O_l) due to greater moment (2475 ft·lb versus 84 ft·lb resistance, Q_p). The second joint of pipe would then push the first joint at an outward angle with a vectored vertical force (resultant). The mathematical analysis suddenly becomes more complex. The type of connection would further influence this motion [i.e., a slip fit coupling would permit a hinge effect, whereas a solid (welded) connection would simulate a single rigid pipe]. For the former, the pipe configuration would be quite disjointed in appearance, with each joint assuming a different direction (see Fig. 5.12*b*). The welded joint pipe, behaving as a single long pile, would assume a profile similar to an elongated S (see Fig. 5.12*c*). (Both configurations assume a homogeneous soil with no obstructions.) As additional joints are driven (pile lengthens), the force Q_L becomes less significant, and the driving force Q_J and the soil's resistance Q_R become controlling factors. Accordingly the complexities of the process become formidable issues. The degree of deflection is generally indeterminate. This is complicated even further by the pier's resistance to bending. However, it's a certainty that the piles will deviate from vertical to the extent that all vertical support capacity is

---

[*]Dr. Stephenson C.E. Professor, University of Missouri-Rolla questions the "wicking." However, the problems associated with anchor heave is well documented in the DFW area.

(a)

(b)

FIGURE 5.9 *(a), (b)*   Minipile failure.

(c)

FIGURE 5.9 (c)   The obvious and serious rake in yet another instance of failed steel minipiles. This scene shows two minipiles exposed by a backhoe.

threatened if not lost. This final failure may be delayed for some period of time (see Fig. 5.9).

The author gives special thanks to Dr. Richard Stephenson, Dept. of C.E., University of Missouri-Rolla and to Dr. S. N. Endley and Dr. K. Mohan Vennalaganti, PSI Inc., Houston, for their valuable input on the pile behavior under eccentric loading.

FIGURE 5.10   One of the problems that can occur as a result of using the perimeter beam as the resistance to drive the steel minipile. Not only have the bolts sheared (losing all support for the structure), but the foundation beam has also been seriously sheared.

5.4.1.2   Costs.   Aside from the other problems inherent to minipiles, their cost is inordinate—often 1.8 to 4 times the price for a conventional 12-in (30-mm) -diameter drilled concrete pier. Perhaps the *only* saving grace lies with the fact that exterior (perimeter) piles can be installed more quickly and with somewhat less damage to the landscaping. However, bear in mind that the cost differential between a single piling and a concrete pier is sufficient to purchase a pick-up truck load of landscape plants. Is the idea of shortening the time for repairs by a day or so worth the sacrifice? Another *seeming* advantage might be the fact that some of the minipile contractors offer *lifetime* warranties. Whenever you are offered one of these, solicit legal advice. You may find words but questionable protection (see Chap. 10).

The cost for the steel-driven pilings varies significantly. This variation is brought about by (1) differences in technique, (2) franchise or royalty fees, (3) differences in excavation costs (deeper perimeter beams or footings require increased excavation, which in some cases might extend to depths at which shoring is required), (4) geographic locations, (5) local codes, and (6) closer spacing [nominally the smaller-diameter underpins are located on 3- to 6-ft (0.9- to 1.8-m) centers]. The range of the cost spans $450 to $1500 (the latter, upside cost was supplied by *PBF Magazine,* May 15, 1997, in the article "How Deep Will They Go?").

5.4.2 Ultraslim Concrete Piers

Whether hydraulically driven or poured in place, the ultraslim concrete piers are also not generally reliable as underpinning options.[16,17] This is particularly true when they are used in areas with expansive soil.

5.4.2.1 Hydraulically Driven Concrete Cylinders. The process of hydraulically driven concrete cylinders, nominally 12 in (30 cm) in length and 4 to 6 (10 to 15 cm) in diameter, (whether strung on a central cable or

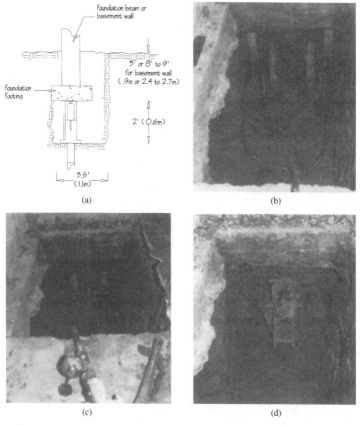

(a)

(b)

(c)

(d)

FIGURE 5.11 Concentric pile technique. (*a*) Drawing of equipment in place; (*b*) photo of equipment in place; (*c*) photo of beam being raised; (*d*) beam and pier locked in place.

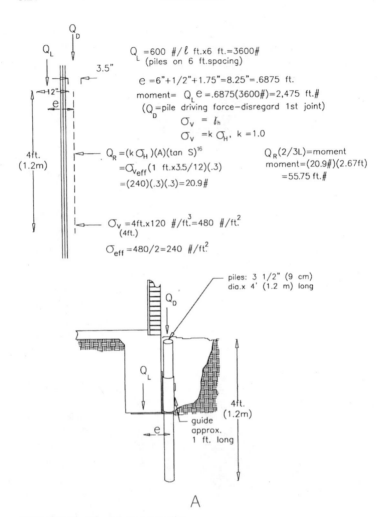

$Q_L = 600 \ \#/\ell \ \text{ft.} \times 6 \ \text{ft.} = 3600\#$
(piles on 6 ft.spacing)

$e = 6" + 1/2" + 1.75" = 8.25" = .6875 \ \text{ft.}$

moment $= Q_L \, e = .6875(3600\#) = 2,475 \ \text{ft.}\#$
(Q_D = pile driving force − disregard 1st joint)

$\sigma_V = \ell_h$
$\sigma_V = k \, \sigma_H, \ k = 1.0$

$Q_R = (k \, \sigma_H)(A)(\tan S)^{16}$
$= \sigma_{Veff}(1 \ \text{ft.} \times 3.5/12)(.3)$
$= (240)(.3)(.3) = 20.9\#$

$Q_R(2/3L) = \text{moment}$
moment $= (20.9\#)(2.67\text{ft})$
$= 55.75 \ \text{ft.}\#$

$\sigma_V = 4\text{ft.} \times 120 \ \#/\text{ft.}^3 = 480 \ \#/\text{ft.}^2$
(4ft.)

$\sigma_{eff} = 480/2 = 240 \ \#/\text{ft.}^2$

piles: 3 1/2" (9 cm)
dia.x 4' (1.2 m) long

Q_D

Q_L

e

guide
approx.
1 ft. long

4ft.
(1.2m)

A

FIGURE 5.12 Eccentrically driven piles.

merely stacked), is prone to failure in lateral stress. If it becomes a concern, they also have virtually no resistance to tensile stress, particularly in those instances where the cable is not stressed. Figure 5.13 is an artist's rendering of this pile system. When the method was first introduced, the procedure was to drive the concrete test cylinders into the ground beneath the perimeter

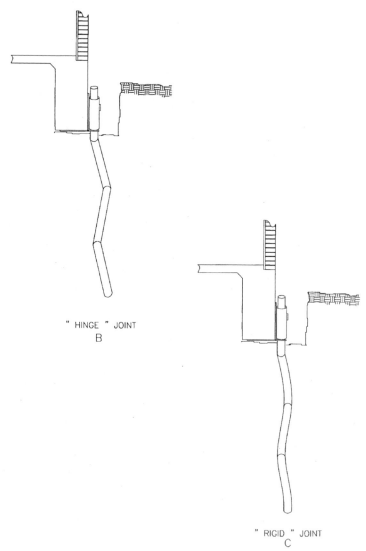

" HINGE " JOINT
B

" RIGID " JOINT
C

FIGURE 5.12 (*Continued*) Eccentrically driven piles.

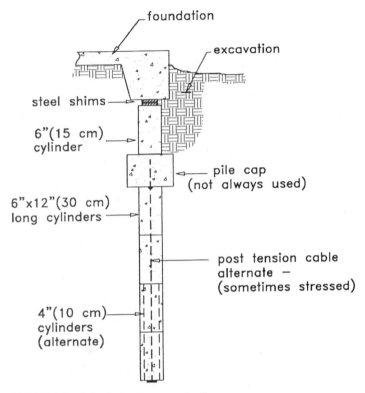

foundation

excavation

steel shims

6"(15 cm) cylinder

pile cap
(not always used)

6"x12"(30 cm)
long cylinders

post tension cable
alternate –
(sometimes stressed)

4"(10 cm)
cylinders
(alternate)

FIGURE 5.13 Hydraulically driven concrete pilings.

beam. The cylinders were stacked on top of each other and the driven length of cylinders was referred to as a "pier."

This poses a real stretch of the imagination. Next, seemingly to overcome the criticism of alignment and total lack of tensile or lateral resistance to stress, the cylinders were strung on a single cable $\frac{3}{8}$-in (0.9-cm) post-tension cable. This is somewhat comparable to digging a 6-in (15-cm) -diameter fence posthole to some depth, inserting a single #3 ($\frac{3}{8}$ in, or 0.9 cm) rebar, pouring concrete, and representing the result as a "structural pier." The difference is due to the singular fact that the cable provides more resistance to shear than the rebar. A 270-k $\frac{3}{8}$-in diameter cable provides shear resistance of about 22,950 lb—provided it is tensioned. A #3 rebar (40,000 psi) provides a resistance of only 4400 lb. Could anyone be convinced that this would work? Let's bring back the Montana ocean-front property. In fact, this method appears to have little market value under any circumstances.

The cost to the consumer for these underpins is reportedly in the range of $225 to $350 each. However, due to the closer spacing, the effective "job" cost will increase proportionately.

5.4.2.2 Ultraslim Piers (Poured Concrete).

Usually 8 in (10 to 20 cm) in diameter, and nominally 5 to 6 ft (1.5 to 1.8 m) on centers, the drilled ultraslim piers suffer problems due to stress failures (tensile and lateral). Adequate steel reinforcing can provide some desired resistance to these stresses. However, the small diameter makes the proper placement of both concrete [$1^1/2$ in (3.75 cm) minus aggregate] and steel most difficult.[16,17] Figure 5.14 illustrates the problem. With the steel in place, the clearance is only 1.875 in (4.76 cm). (This can vary some with the manner in which the steel is caged.) A safe passage for $1^1/2$-in (3.75-cm) rock requires a minimum clearance of 4.5 in (11 cm). [Proper clearance for concrete placement is generally assumed to be 3 times the diameter of the largest-size aggregate. In the case of concrete with $1^1/2$-in rock minus aggregate, the desired clearance would be 3×1.5 in, or 4.5 in (11 cm).] Concrete technology also suggests a 3-in (7.6-cm) coverage of concrete between steel and steel or steel and form. The three #3s practically preclude the placement of regular hard rock concrete. Concrete strength must be sacrificed by either opting for a pea gravel concrete or decentralized steel. Neither option is acceptable. [*Conventional* piers often use as many as four #3s (0.95 cm), #4s (1.27 cm), or #5s (1.6 cm). With these, the steel can be tied in a 6-in (15-cm) cage, which allows ample clearance for concrete.]

The 8-in (20-cm) -diameter piers are often used in tandem to presumably offer bearing support comparable with conventional 12-in (30-cm) -diameter concrete piers. In linear strength calculations, a single 12-in (30-cm)- diameter concrete pier (without steel reinforcement) is equivalent to 2.25- to 8-in (20-cm) -diameter piers. The big difference between the two is noted when (1) the piers are subjected to eccentric loading or lateral stress or (2) steel rebar interferes with concrete placement.

Dual ultraslim piers have been suggested. The geometrics of the tandem installation causes the individual piers to be raked—often well in excess of 20°. The recommended rake for any load-bearing pier is 10 to 15°.[16,17,91] Also, the dual piers do not overcome the obstacles cited for the single ultra- slim pier. Refer to Sec. 5.4.3 and Figs. 5.14 and 5.15. In the case of slab foun- dations, proper mudjacking (following the underpinning) might somewhat overcome the inadequacies of the ultraslim minipiles.

5.4.2.3 Cost.

As opposed to the steel minipiles, either of the ultraslim concrete piers is less costly than conventional concrete piers or spread foot- ings, often in the range of $200 to $275 each. A contractor who charges $200 each for the ultraslim drilled pier should enjoy much higher profit than a con-

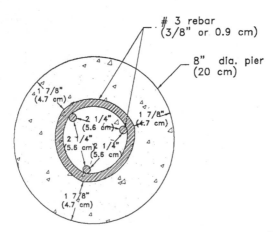

concrete clearance between rebar and 8" dia. pier:

    preferred clearance is usually specified
    as 3" or 3 times maximum diameter
    of aggregate

FIGURE 5.14   Poured concrete slim pier.

tractor who, under similar conditions, installs a 12-in (0.3-m) pier for $300
to $350. The cost to put in an 8-in (20-cm) pier is less than one-half the cost
necessary for the larger pier.

5.4.3   Combination Spreadfootings and Drilled Pier

On occasion, piers or pilings are used in conjunction with the spreadfooting
(haunch). Here the theory is to utilize the best support features of each design
in the hope of achieving a synergistic effect. (Also, as a practical matter, a
haunch is always necessary to provide a base from which to raise the beam.)
In practice, the goal is not always attained. When soil conditions dictate the
spreadfooting, the pier provides little, if any, added benefit. The integration
of the deep pier as part of the spreadfooting, generally, has no deleterious features, provided (1) the cross-sectional area of the footing is not diminished,
(2) the pier does not penetrate a highly expansive substratum that has access
to water, (3) the diameter of the pier is at least 10 in (25 cm), and (4) the pier
is not raked to an excessive degree (see Fig. 5.15). Water in contact with the
CH (clay of high plasticity) lay at a 5- to 7½-ft (1.5- to 2.3-m) depth could
cause the piers to heave, whereas the spreadfooting would be stable. Refer to
"The Effects of Soil Moisture on the Behavior or Residential Foundations in

Active Soils," R. W. Brown and C. H. Smith, *Texas Contractor,* May 1980. Certain steps are available to avoid or minimize friction (upheaval) in the design of shallow piers. These efforts include (1) using the "needle" or "slim" pier or piling (reduced surface area), (2) belling the pier bottoms (usually effective with conventional diameter shafts), and (3) placing a friction-reducing membrane between the pier and the sidewall of the hole.

When dual piers are used, several alterations are required. First, the pier diameters are generally limited to about 8 in (20 cm). Second, the dual piers must be raked (deviated from vertical). This rake should never exceed 12°, but in practice it often exceeds 20°.[37] The small-diameter pier introduces all the drawbacks described in Sec. 5.4.2.2. The cost for these piers is equivalent to, or exceeds, that for the conventional 12-in (0.3-m) concrete piers, but obviously they are less effective.

5.4.4 Hydropiers

The so-called hydropier is too weak in both theory and practice to merit discussion. Maintaining a constant level of soil moisture is certainly a beneficial control for expansive soil movement. However, water injection alone is not

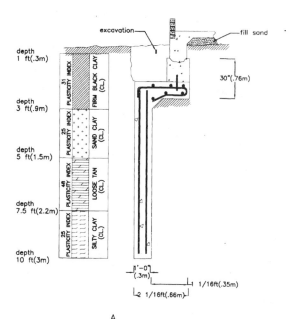

A

FIGURE 5.15 (*a*) Conventional pier.

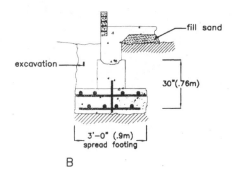

excavation

fill sand

30"(.76m)

3'-0" (.9m)
spread footing

B

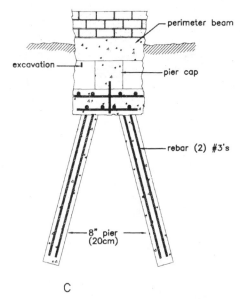

perimeter beam

excavation

pier cap

rebar (2) #3's

8" pier
(20cm)

C

FIGURE 5.15 (*Continued*) (*b*) Spreadfooting, (*c*) dual minipier.

likely to provide any degree of beneficial leveling and can cause serious damage.[3] This method does not really produce a "pier"; hence the label is misleading. Figure 5.16 shows the placement and development of a typical hydropier. The principal claims seem to be

1. The system continuously supplies water to the soil, thus preventing settlement.

2. Another company using an identical process advertises "uniform foundation raising" of up to 3 in (7.5 cm).

3. Some users claim that the vertical weep hose delivers water that expands the clay adjacent to the hose, thus creating a "hydro pier."

4. Others claim that the expansion of the clay adjacent to the hose constricts the hose to the extent that at some point water flow is shut off.

No one can take this seriously.

5.4.5 Underpinning an Interior Slab versus Mudjacking

Figure 5.17 is an engineer's rendering for restoring an interior slab floor to (or near) original grade. The 28 "dots" represent pier locations to be installed by breaking out sections of the slab floors. (It is the author's opinion that a slab foundation should be broken only in extremely rare occasions, such as for raising an interior fireplace.) The installation of the interior piers (1) threatens the structural integrity of the slab, (2) creates a horrendous mess, and (3) is inordinately expensive. On this particular job, the owner alluded to a bid of "well over $12,000" to install the piers/pilings and fill voids. A bid

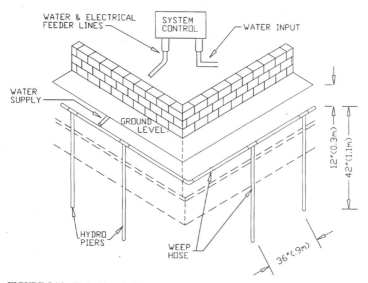

FIGURE 5.16 Typical installation—hydropiers.

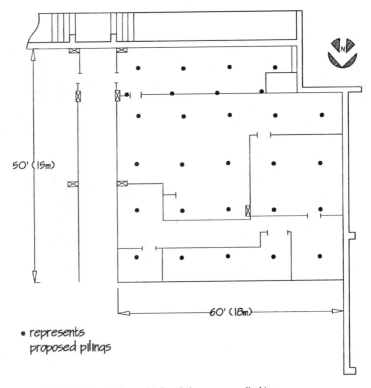

50' (15m)

60' (18m)

• represents
  proposed pilings

FIGURE 5.17 Underpinning a slab foundation versus mudjacking.

for competent mudjacking was less than $4000. The job was mudjacked, the owner was fully satisfied, and there was little inconvenience to the tenants and no significant damage to the floor slab.

5.5 SUMMARY

Regardless of the type of underpinning support, two precautions must be exercised. First, concrete should be poured into the pier shafts or pads as quickly after excavation as possible. Ideally, concrete would be poured the same day as excavation. Second, no steel or wood should be exposed below grade either as shim or pier materials. Exposure to water and soil will corrode the steel or rot the wood within an unusually short period of time. Third, the pier cap should be poured in place to ensure intimate contact between the pier and the irregular bottom of the perimeter beam. Shims or flat surfaces in contact with the irregular surface of the beam will result in either immediate damage to the concrete beam or subsequent resettlement due to the same effect over a period of time, both due to crushing of concrete protrusions. Fourth, concrete placed into shafts deeper than about 15 ft (4.5 m) should be tremmied.

Also, bear in mind that none of the underpinning techniques attempt to "fix" the foundation to prevent upward movement. This is by design to avoid subsequent, uncontrolled damage to the foundation and emphasizes the fact that, if shallow expansive soils are subjected to sufficient water, the soil expansion will raise the foundation off the supports.

5.6 CONCRETE DESIGN TABLES

Tables 5.1 through 5.4 give a ready correlation for slump, concrete composition, and relative strengths, as well as weight and size of rebar. These tables should be self-explanatory. Table 5.5 gives the amount of concrete required to pour at various diameters.

TABLE 5.1 Recommended Slumps for Concrete

	Slump, in (cm)	
Types of structure	Minimum	Maximum
Massive sections, pavements, and floor laid on ground	1 (2.54)	4 (10.16)
Heavy slabs, beams, or walls; tank walls; posts	3 (7.62)	6 (15.24)
Thin walls and columns; ordinary slabs or beams; vases and garden furniture	4 (10.16)	8 (20.32)

TABLE 5.2 Mixture for 1 yd³ (0.76 m³) of 3000-lb/in² (21-MPa) Concrete

Material	Amount/sack of cement	Total amount/yd
Cement (5 sacks)	94 lb (42.5 kg)	470 lb (212 kg)
Sand	314 lb (142.5 kg)	1570 lb (712 kg)
Coarse aggregate	345 lb (157 kg)	1725 lb (784 kg)
Water	7 gal (max.) (26.5 l)	35 gal (132.5 l)

TABLE 5.3 Reinforcement Grades and Strength

	Min. yield strength f_y, lb/in^2 (MPa)	Ultimate strength f_u, lb/in^2 (MPa)
Billet steel grade		
40	40,000 (276)	70,000 (483)
60	60,000 (414)	90,000 (620)
75	75,000 (517)	100,000 (690)
Rail steel grade		
50	50,000 (345)	80,000 (552)
60	60,000 (414)	90,000 (620)
Deformed wire		
Reinforced	75,000 (517)	85,000 (586)
Fabric	70,000 (483)	80,000 (552)
Cold-drawn wire		
Reinforced	70,000 (483)	80,000 (552)
Fabric	65,000 (448)	75,000 (517)

The following example calculates the amount of concrete that should be ordered to pour 15 piers, 12 in (0.3 m) in diameter, to a depth of 20 ft (6 m):

$$15 \times 0.029 \text{ yd}^3/\text{ft} \times 20 \text{ ft} = 8.7 \text{ yd}^3 \ (6.6 \text{ m}^3)$$

The following calculates the amount of concrete that should be ordered to pour 15 piers, 18 in (0.45 m) in diameter, to a total depth of 18 ft (5.4 m):

$$15 \times 0.065 \text{ yd}^3/\text{ft} \times 18 \text{ ft} = 17.55 \text{ yd}^3 \ (13.3 \text{ m}^3)$$

$$\text{Extra for bell} = 15 \times 0.255 = 3.85 \text{ yd}^3 \ (2.9 \text{ m}^3)$$

$$\text{Concrete required} = 17.55 + 3.85 = 21.4 \text{ yd}^3 \ (16.26 \text{ m}^3)$$

Table 5.7 assists with the cursory selection of safe bearing areas required to support various loads on soils of a given unconfined compressive strength. For example, for a soil with an unconfined compressive strength of 1000 lb/ft^2 (157 kN/m^2) and an intended load of 4000 lb (1818 kg), none of the conventional piers would safely accommodate this load in end-bearing alone. However, a 12-in (3-m) pier with a 2-ft^2 (0.372 m^2) haunch would carry the load. Again, this approach is intended for screening purposes only. Any concerns should be resolved by a complete mathematical analysis based on the specific and complete project specifications. Suggestions provided by the Table 5.6 are generally quite conservative.

At the end of the day, the structural load must be accommodated by the soil. The foundation and any underpinning merely represent devices used to

TABLE 5.4 Weight, Area, and Perimeter of Individual Bars

Bar dimension #	Unit weight, lb/ft (kg/m)	Diameter d, in (cm)	Cross-section area (CSA), in² (cm²)	Perimeter, in. (cm)
2	0.167 (0.249)	0.250 (0.635)	0.05 (0.32)	0.786 (2.0)
3	0.376 (0.560)	0.375 (0.95)	0.11 (0.71)	1.178 (2.99)
4	0.668 (0.995)	0.500 (1.27)	0.20 (1.29)	1.571 (3.99)
5	1.043 (1.560)	0.625 (1.59)	0.31 (2.0)	1.963 (4.99)
6	1.502 (2.24)	0.750 (1.9)	0.44 (2.84)	2.356 (5.98)
7	2.044 (3.05)	0.875 (2.22)	0.60 (3.87)	2.749 (6.98)
8	2.670 (3.99)	1.00 (2.54)	0.79 (5.1)	3.142 (7.98)
9	3.400 (5.07)	1.125 (2.86)	1.00 (6.45)	3.544 (9.0)
10	4.303 (6.41)	1.250 (3.175)	1.27 (8.19)	3.990 (10.13)
11	5.313 (7.92)	1.375 (3.49)	1.56 (10.06)	4.430 (11.25)
14	7.65 (11.40)	1.750 (4.45)	2.25 (14.52)	5.32 (13.51)
18	13.60 (20.26)	2.250 (5.715)	4.00 (25.8)	7.09 (18.0)

5.35

TABLE 5.5 Concrete Piers*

	CSA		Volume per linear foot			Volume 2x bell		
	ft^2	m^2	ft^3	yd^3	m^3	ft^3	yd^3	m^3
10	0.54	0.05	0.54	0.02	0.015			
12	0.785	0.073	0.785	0.029	0.022	1.18	0.044	0.033
14	1.07	0.099	1.07	0.0396	0.03			
16	1.4	0.13	1.4	0.052	0.04			
18	1.77	0.164	1.77	0.065	0.05	6.9	0.255	0.194
24	3.14	0.29	3.14	0.116	0.088	18.84	0.698	0.53

*Pier diameters smaller than 10 in (25 cm) should not be considered for underpinning foundations.

TABLE 5.6 Post-Tension Cables

Strand size f_{pu}, in-ksi	Strand area, in^2	0.7 f_{pu} Aps, kips	Average prestress loss,* kips	Effective prestress, kips
3/8-270	0.085	16.10	1.3	14.8
7/16-270	0.115	21.70	1.7	20.0
1/2-270	0.153	28.90	2.3	26.6

* Assumed prestressed losses of 15 ksi; actual losses should be calculated as per Sec. 6.6.[90]
Note that the effective strand CSA is less than that for a comparable deformed bar. For example, the CSA of a #3 rebar is 0.11 in, whereas that for the 3/8-in cable is 0.085 in. The effective area is the actual combined cross-sectional areas of the individual strands making up the cable. The term f_{pu} is equivalent to f_u in Table 5.3. In practice the process of tensioning the cables is associated with several losses in tensile strength. More on this can be found in the PTI manual.[90]

distribute structural loads to perfect a soil advantage. A simple review of the weight-bearing capacity of a typical soil is shown in the following analysis.

Figure 5.18 depicts a typical poured concrete pier beneath the perimeter beam of a slab foundation (see also Fig. 5.1).

Assume a uniform load on the perimeter beam equivalent to 600 lb per linear foot. The piers are to be located on 8-ft centers. The structural weight distributed to each pier location would then be Q_w = 600 lb/1 ft × 8 ft = 4800 lb. The piers are assumed to be vertical.

The resistance to this load provided by the soil can be summarized as follows:

Pier end-bearing only; assume q_u = 3 tsf, $\pi/4$ = 0.785 and A = 0.785D^2:

$$Q_{EB} = 0.785 \text{ ft}^2 \times 3 \text{ tsf}$$
$$= 2.36 \text{ tons} = 4720 \text{ lb}$$

Resistance provided by haunch:

$$Q_H = 2.5 \text{ tsf} \times (5.0 - 0.985) \text{ ft}^2$$
$$= 2.5 \text{ tsf} \times 4.215 \text{ ft}^2 = 10.54 \text{ tons}$$
$$= 21,080 \text{ lb}$$

Weight capacity of perimeter beam in full contact with soil or subsequent to mudjacking:

$$Q_s = (0.833 \text{ ft} \times 1 \text{ ft}) \, 2 \text{ tsf}$$
$$= 1.666 \text{ tons per linear foot beam}$$

Assuming an 8-ft increment,

$$Q_x = \frac{3332 \text{ lb}}{1 \text{ ft}} \times (8-1) \text{ ft}$$
$$= 3332 \, (7) = 23,324 \text{ lb}$$

At this point the maximum load support for the soil in question is

$$Q_{max} = Q_{EB} + Q_H + Q_S$$
$$= 4720 + 21,080 + 23,324$$
$$= 49,124 \text{ lb}$$

The factor of safety under the prevailing conditions would be

$$SF = \frac{49,124}{4800} = 10.23$$

If the pier skin friction were to be considered, the soil bearing capacity would increase by

$$Q_{friction} = (K \, \bar{\sigma}_v \tan S) \, A_{surface}$$

Refer to references 12 and 15–17. Assume $\gamma = 125 \text{ lb/ft}^3$, $\tan S = 0.45$[12,15–17]

$$Q_{friction} = (1 \times 7 \times 125 + 12 \times 125/2 \, (0.45) \, (5 \text{ ft}) \, (\Pi) \, (1 \text{ ft})$$

$$= \left(\frac{875 + 1500}{2} \text{ lb/ft}^2 \right) (0.45) \, (15.75 \text{ ft}^2)$$

$$= (1,187.5 \text{ lb}) \, (0.45) \, (15.75)$$

$$= 8,416 \text{ lb}$$

TABLE 5.7 Maximum Load-Bearing Capacity (Q), lb (N)*

Columns under the spanning header **Unconfined Compression Strength of Soil (q_u)†, lb/ft² (N/m²)†**

Support	Bearing area, ft² (m²)	1000 (4448)	2000 (8869)	3k (13,344)	4k (17,792)	6k (26,688)	8k (35,584)
3 in dia.	0.05 (0.0045)	50 (20)	100 (40)	150 (60)	200 (80)	300 (120)	400 (160)
4 in dia.	0.09 (0.008)	90 (35.6)	180 (71)	270 (107)	360 (142)	540 (214)	720 (284)
6 in dia.	0.20 (0.018)	200 (80)	400 (160)	600 (240)	800 (320)	1200 (480)	1600 (640)
8 in dia.	0.35 (0.032)	350 (142)	700 (284)	1050 (426)	1400 (568)	2100 (852)	2800 (1136)
10 in dia.	0.55 (0.05)	550 (222)	1100 (444)	1650 (666)	2200 (888)	3300 (1332)	4400 (1776)
12 in dia.	0.785 (0.070)	785 (315)	1570 (630)	2355 (945)	3140 (1260)	4710 (1890)	6280 (2520)
1 ft²	1.0 (0.090)	1000 (414)	2000 (826)	3000 (1242)	4000 (1626)	6000 (2484)	8000 (3312)
16 in dia.	1.4 (0.13)	1400 (578)	2800 (1156)	4200 (1734)	5600 (2312)	8400 (3468)	11,200 (4624)
18 in dia.	1.76 (0.16)	1760 (711)	3520 (1423)	5280 (2133)	7040 (2847)	10,560 (4266)	14,080 (5693)
2 ft²	4 (0.36)	4000 (1601)	8000 (3200)	12k (4800)	16k (6400)	24k (9600)	32,000
2.5 ft²	6.25 (0.56)	6250 (2500)	12.5k (5000)	18.75k (7500)	25k (10,000)	37.5k (15,000)	
3 ft²	9.0 (0.81)	9000 (3603)	18k (7266)	27k (10,809)			

*The values given in the table do not include a margin of safety. The indicated total safe load should be divided by the appropriate safety factor to give the safe design load. For example, a soil with a q_u of 2000 lb/ft² would require a minimum bearing area (CSA) of 1 ft² to handle a weight load of 2000 lb, again without regard to any safety factor.

†k = 1000 lb.

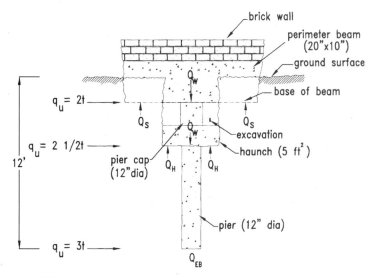

FIGURE 5.18   Soil load capacity.

Note: Omit the top 7 ft of pier depth for friction calculations.[12,15–17] The total combined soil bearing now becomes 49,124 + 8416 or 57,540 lb. The new safety factor becomes 54,570/4800 = 512. As an aside, now consider a 6-in-diameter pressed pile (see Fig. 5.13). There is no haunch to consider, only the pier end-bearing: $q_u = 3$ ts

$$Q_{EB} = (0.785) \left(\frac{6 \text{ in}}{12 \text{ in}} \right)^2 \times 3 \text{ tsf}$$

$$= 0.196 \text{ ft}^2 \times 3 \text{ tsf} = 0.589 \text{ tons}$$

$$= 1177.5 \text{ lb}$$

This pier (soil actually) would not carry the structural load of 4800 lb or, for that matter, 3000 lb if the pier spacing were reduced to 5 ft. The nature of the pressed pile is such that no design benefit can be attributed to skin friction. Probably the pier could function if enhanced by the load capacity provided by the perimeter beam. Proper mudjacking to ensure intimate soil-beam contact would be a must. Therefore,

$$Q_S + Q_{EB} = 23,324 + 1177.5$$

$$= 24,501$$

Under these conditions, the factor of safety could become

$$SF = \frac{24,501}{4800} = 5.1$$

The end-bearing capacity of an 8-in diameter pier under the same conditions would be

$$Q_{EB} = .785 \left(\frac{8}{12}\right)^2 \text{ ft}^2 \times 3 \text{ tsf}$$

$$= 0.35 \times 3 = 1.05 \text{ tons}$$

$$= 2093.3 \text{ lb}$$

This pier would not carry the structural load without help. The usual 4-ft^2 haunch would provide some help. For example,

$$Q_H = 2.5 \text{ tsf} \times (4 - 0.35) \text{ ft}^2$$

$$= 2.5 \text{ tsf } (3.65) \text{ ft}^2 = 0.125 \text{ tons}$$

$$= 18,250 \text{ lb}$$

Added to the end-bearing the total unit bearing capacity becomes

$$Q_{EB} + Q_H = 18,250 + 2093$$

$$= 20,343 \text{ lb}$$

$$SF = \frac{20,343}{4800} = 4.2$$

The skin friction would also provide additional load capacity.

The same or a similar analysis can be used to screen potential designs for underpins. Sometimes simple variations in design features can be incorporated to enable the desired factor of safety.

CHAPTER 6
BASEMENT OR FOUNDATION WALL REPAIR

6.1 INTRODUCTION

The remedial approaches to basements do not fit in with "normal" foundation repair procedures. This introduces a special-case scenario that is of concern only in certain locales. Basement construction is desirable in instances where (1) the source for heat needs to be lower in elevation than the living area; heat rises; (2) land costs are high, making it less expensive to excavate than to spread out; and (3) a frost line or permafrost must be considered. In the south, few basements are built or have been built during the last 60 years. Therefore, the author's experience with basement repair is limited. The following examples were taken from this limited exposure plus repair procedures designed by structural engineers (P.E.s) over the country.

6.2 TYPICAL APPROACHES TO BASEMENT REPAIRS

The first repair example involves the construction of a new wall inside the original wall. This is sometimes referred to as a "sister" wall.[13] The structural load is ultimately transferred to this addition. No effort is made to plumb or reinforce the existing basement foundation wall (see Fig. 6.1a). The sequence of construction is essentially as follows:

6.1

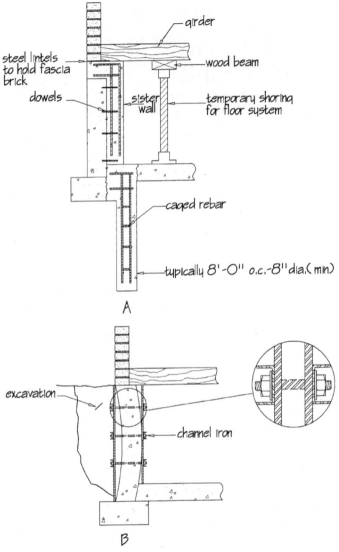

girder

steel lintels
to hold fascia
brick

wood beam

dowels

sister
wall

temporary shoring
for floor system

caged rebar

typically 8'-0" o.c.-8" dia. (min)

A

excavation

channel iron

B

FIGURE 6.1 Basement/foundation wall repair. (*a*) Failure in foundation (basement) wall—transfer of load; (*b*) failure in foundation wall, plumbing desired.

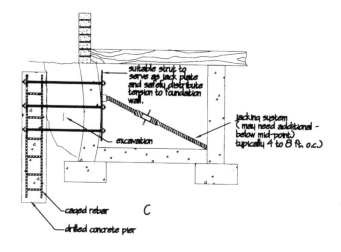

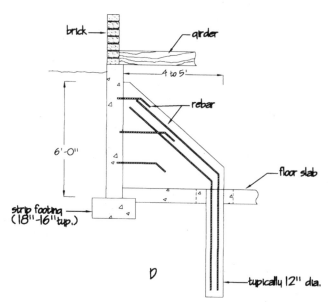

FIGURE 6.1 *(Continued)* Basement/foundation wall repair. (*c*) Failure of basement wall, plumbing desired; (*d*) knee brace to control rotation.

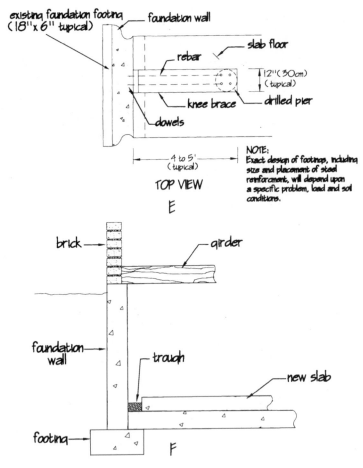

FIGURE 6.1 *(Continued)* Basement/foundation wall repair. *(e, f)* Water control.

1. Shore existing floor joists to lessen or remove the structural load from the defective wall.

2. Sandblast or scarify surface of existing wall to remove laitance (loose concrete material) and provide an improved bonding surface.

3. Break out the floor slab (and protruding strip footing where applicable) to prepare for installation of supporting concrete piers and beam (where

applicable). Typically, for a wall height less than about 10 ft (3 m) with construction loads less than about 1500 lb per linear foot (2268 kg/m), the piers could be 10 to 12 in in diameter (25 to 30 cm) spaced on approximately 8-ft (2.4-m) centers.

4. Place steel lintels across keyways to hold fascia brick. An alternative is to remove brick.

5. Break out concrete or remove sufficient concrete block to create keyways of approximately 1×2 ft $(0.3 \times 0.6$ m), 6 ft (1.8 m) on centers (OC).

6. Drill dowel holes into existing foundation/basement wall. Often the rebars would be 7s ($^7/_8$ in, or 2.2 cm) spaced to create a pattern on the order of 12 to 18 ft^2 (1.0 to 1.5 m^2) in area.

7. Place steel and pour piers and beam.

8. Place steel and pour sister wall. Often the concrete is placed through the keyways by concrete pumps.

Note: This and the following examples are meant to be general. Specific loads and job conditions will dictate the size reinforcement and spacing of the support members. Also, virtually every installation will benefit from some type of waterproofing.

Figure 6.1*b* presents yet another problem. This approach permits some plumbing of the defective wall and is generally more effective with masonry wall construction than with concrete. The wall materials should suffer little or no deterioration. The general procedure for this installation is

1. Excavate fill adjacent to wall exterior.

2. Drill holes through basement/foundation wall to facilitate take-up bolts. Generally, the bolts would be 1 to $1^1/_2$ in (2.54 to 3.8 cm) in diameter, located at least three to a tier and 3 to 6 ft (0.9 to 1.8) OC.

3. Place channel iron on each side of wall. Install bolts and tighten.

4. Waterproof wall as necessary and backfill with a gravel hydrostatic drain system.

5. As with virtually all attempts to plumb a failed basement/foundation wall, it is often desirable to supplement the system with additional force to push the wall into plumb (see Fig. 6.1*c*).

Another technique, more suited to actually plumbing a concrete basement/foundation wall, is depicted in Fig. 6.1*c*. This technique uses an opposing wall to secure the jacking system that is utilized to plumb or align the basement/foundation wall. Depending on such factors as existing wall design, load conditions, and degree of rotation, a second battery of jacks may be required. Typically each battery of jacks is placed 4 to 8 ft (1.2 to 2.4 m) apart. A typical sequence for the installation of this system is

1. Drill holes for dywidag (or similar) bars, $1^1/2$ to 2 in (3.8 to 5.0 cm) in diameter, through basement/foundation wall. Locate bar in external drilled pier shaft. Holes are typically three to each pier, 4 to 8 ft (1.2 to 2.4 m) OC.

2. Drill and pour the external concrete piers.

3. Excavate behind the existing wall. (Alternatively, the backfill could be excavated prior to placement of the pier. This would necessitate forming the piers but would allow the strip footing to be broken out at pier locations. Removal of the protruding strip footing would allow contact of the pier to existing wall with vertical alignment.)

4. Place channel iron and commence jacking operation.

5. Waterproof wall as required and backfill with suitable gravel hydrostatic drain.

The knee brace is yet another approach to retrofit basement/foundation walls. This method is fairly simple and adequate to sustain wall rotation. This method is not intended to accomplish any degree of vertical alignment of the *existing wall.* Figure 6.1*d* depicts the design of a typical knee brace. Depending on specific conditions, a second pier may be required immediately adjacent to the existing wall. However, for lightly loaded conditions, a schedule of dowels will prevent any slip between the wall and brace. Considering normal basement heights [less than 10 ft (3 m)] and lightly loaded conditions, the placement of the braces might be 6 to 10 ft (1.8 to 3 m) OC.

Note: The knee brace is also frequently used to control outward rotation of foundation walls in deck-high construction. The primary limitation would be instances where the defective wall is situated on a "zero lot line." When the foundation walls are 5 ft (1.5 m) high or less and the wall structure permits, the placement of the supports might be as far apart as 10 to 20 ft (3 to 6 m). For certain minor problems, particularly those associated with lightly loaded conditions, an adequate restoration approach could be to utilize the jacking system illustrated in Fig. 6.1*a* to raise and level the floor joist system. Permanent supports would consist of supplemental beams (wood as a rule) supported atop lally columns. Generally, the columns [often 4- to 5-in (10- to 15-cm) steel pipe] are fitted with steel plates top and bottom. Frame basement walls are particularly suited to this option. Job conditions control the number, design, and placement of these supports.

6.3   HADITE BLOCK WALLS

Other relatively minor problems sometimes involve hollow concrete block (Hadite) walls. If the intent is to strengthen the wall or shut off minor water seepage, the problem might be addressed by filling the blocks with a concrete mix. Normally, at least two rows of injection holes are drilled through the inside of a concrete block wall. Typically, the lowest row would be about 4 ft (1.2 m) above the floor and a second row near the top of the wall. Adjacent (lateral) holes are used to ensure complete penetration of all voids. Initially, the holes might be approximately 4 ft (1.2 cm) OC. If any question arises concerning filling all voids, intermediate holes can be drilled. The lowest row of holes is normally injected first and may or may not be allowed to attain initial set before the top row is pumped. The stage pumping reduces the hydrostatic load on the lower courses of block.

6.4   BASEMENT WATER INFILTRATION

With persistent water infiltration, the solution shown in Fig. 6.1*e* might be acceptable. The water that penetrates the wall is collected in the trough at the perimeter of the floor slab. Water is then transported to an adequate sump or drain. A slight variation with this is to (1) construct a sump in the floor slab and (2) build a screed floor over the concrete slab. The sump accumulates the water and pumps it to a drain. The wood screed floor remains dry. The screed members should be rot-resistant (i.e., redwood, cypress, cedar, or chemically treated pine). Neither of these options address structural concerns per se. The expressed intent is to control unwanted water.

6.5   CONCLUSION

When basement wall failures occur in a vertical direction, underpinning and mudjacking are employed as discussed in Chaps. 3 and 5. To avoid the extensive excavation, the repair work is often performed from inside the basement.

There are many other options to correct problems with basement/foundation walls; however, the foregoing should provide a sense of direction.

CHAPTER 7
SOIL STABILIZATION

7.1 INTRODUCTION

Soil stabilization is a procedure for improving natural soil properties to provide more adequate resistance to erosion, water seepage, and other environmental forces and more loading capacity. In foundation or geotechnical engineering, soil stabilization is divided into two sections (1) mechanical stabilization, which improves the structure of the soil (and consequently the bearing capacity), usually by compaction, and (2) chemical stabilization, which improves the physical properties of the soil by adding or injecting a chemical agent such as sodium silicate, polyacrylamides, lime, fly ash, or bituminous emulsions. Generally, the chemical either reacts with the soil or provides an improved matrix that binds the soil.

In residential foundations, soil stabilization refers not only to improving the compressive strength of shear strength but also to increasing the resistance of the soil to dynamic changes. The latter tends to destroy both the soil's integrity and its structure. Generally, the former relates to stress applied to the soil by the foundation and the latter to the conditions imposed by the environment. Both are related to soil characteristics.

Among the different mechanical stabilization techniques such as preloading (to reduce future settlement), moisture control (to speed up settlement), and compaction or densification (to improve bearing capacity and/or reduce settlement), compaction is generally the least-expensive alternative for residential and commercial buildings. Detailed and specific information can be found in Sec. 5A, *Practical Foundation Engineering Handbook.*[17]

7.2 COMPACTION

Compaction may be accomplished by excavating the surface soil to a depth for residential buildings up to 4 ft (1.3 m) and for commercial buildings up to 6 ft (1.8 m) and then backfilling in controlled layers and compacting the fill to 95 percent compaction. Often the fill material is a replacement type such as some nonplastic (low-plasticity index) soil. In the case of uniform soil (sand), the addition of a fine soil to improve the grain size distribution is advised. The standard compaction tests utilized to evaluate these processes include one of the following:

1. *ASTM D698-70.*   5.5-lb hammer, 12-in drop, $1/30$-ft^3 mold; three layers of soil at 25 blows per layer may be used.
2. *ASTM D-1557-70.*   10-lb hammer, 18-in drop, $1/30$-ft^3 mold; five layers at 25 blows per layer may be used.

Specific details of compaction tests, equipment, and quality control are beyond the scope of this book. The interested reader should consult the references, in particular the *Practical Foundation Engineering Handbook.*[17]

The undrained shear strength of a soil acceptable for a housing site should not be less than 800 lb/ft^2 (38 kPa). In a nonexpansive fill with less than 600-lb/ft^2 (29-kPa) undrained shear strength, compaction, preloading, and grout injection are methods beneficial for improving the soil for light residential construction. For heavier foundation loads, piers or other forms of structural enhancement might be required. This might encompass a more sophisticated foundation design, soil improvement, or both. A discussion of specific soils amenable to mechanical improvement follows.

7.3 GRANULAR SOIL

Granular soils are those composed of particles larger than 0.0075 mm (no. 200 sieve). After proper compaction, most granular soils are modified to give them volume stability and improved frictional resistance. Still, they often retain high permeability. To offset this property the soils might be blended with either a granular material to provide a well-graded soil or a cohesive material to provide bonding or cementation. The latter is intended to provide cohesion under both moist and dry conditions. Silty clay constituents (or other cementitious materials) can also decrease the danger of other instability under either dry or, particularly, wet conditions. Laboratory tests are performed to determine the optimum conditions of compaction and soil modification.

In general, a sandy soil, particularly after stabilization, has a high bearing capacity, but foundations should be placed at a sufficient depth so the soil beneath the loaded member is confined. Foundations in stabilized sand may consist of spreadfootings, mats, piles, or piers, depending principally on the soil density and thickness, cost of soil modification, and imposed loads. In sand deposits (without compaction) spreadfootings are used if the deposit is sufficiently dense to support the loads without excessive settlement. Piers in loose sand deposits should be drilled to firm underlying strata. Skin friction can be considered in the design (or load) criteria for sand or granular soil. The foundation should not be located on sand deposits where relative density is less than 60 percent or a density of 90 percent of the maximum cannot be attained in the soil laboratory. An exception would be where the loose material is completely penetrated (usually by driven piles) as noted above.

7.4 FOUNDATION ON LOESS

Loess is a fine-grained soil formed by the deposit of wind-borne (aeolian) particles. This soil covers 17 percent of the United States (see Fig. I.3). Depths of loess deposits range from 3.3 to 165 ft (1 to 50 m) and depths from 6.6 to 9.9 ft (2 to 3 m) are common. Loess has a specific gravity from about 2.6 to 2.8 and in situ dry density from about 66 to 104 lb/ft^3 (1057 to 1666 kg/m^3) with 90 percent particle size passing a no. 200 sieve. The plastic limits range from 10 to 30 percent. Standard compaction tests produce dry densities of 100 to 110 lb/ft^3 (1602 to 1762 kg/m^3) at optimum moisture content of 12 to 20 percent. As a foundation soil, a loess soil with a density greater than 90 lb/ft^3 (1442 kg/m^3) will often exhibit only limited settlement. The problem with loess is the changing of bearing capacity with saturation. Upon saturation, soil bearing capacity can drop to 90 percent or less than that of the dry loess. Loess is silt, cemented by calcareous materials. The addition of water destroys the cement bonds. (Eroded loess is commonly referred to as a silt deposit.) Loess below the permanent water table is, as one would guess, relatively stable because the water content is constant. Compacted loess can be a satisfactory foundation material for mats and spreadfootings if the density is more than 1.6 g/cm^3 (100 lb/ft^3 or 1600 kg/m^3).

Loess can be stabilized by using lime, lime fly ash, or cement, each followed by compaction. Piers are commonly suggested if in-place specific gravity of the loess is under 1.44 g/cm^3 (90 lb/ft^3, or 1440 kg/m^3). Piles should be driven, or piers drilled, through the loess into the underlying soil layer unless the loess terminates below the water table. Again, loess is often stable within or below the water table.

7.5 FOUNDATIONS ON SANITARY LANDFILL SITES

Sanitary landfill is a euphemism for a garbage dump. Within urban areas, it often becomes necessary to develop a former sanitary landfill for construction. In most instances, the opportunity for soil improvement by normal compaction techniques is denied. Often the preparation of a site for construction requires extensive grouting to fill voids and consolidate the soil. The alternative would be a foundation design that penetrated the fill section into competent material.

Landfill usage for one- or two-story residential buildings, apartments, office buildings, or other light construction can be acceptable if the site is either adequately compacted, stabilized by grout injections, or modified by the addition of lime or cement. This assumes the required bearing capacity of the soil (including the factor of safety) to be within the range of 0.5 to 1.0 tons/ft^2 (4.88 to 9.76 $\times$ 10^3 kg/m^2). In this way, the use of continuous foundations may provide adequate bridging capacity over local soft spots or cavities. Otherwise, piers should be extended to a firm layer underlying the landfill.

7.6 STABILIZING PERMEABLE SOILS

Generally, the concerns with noncohesive, permeable soils involve measures to control or prevent sloughing, improve bearing strengths, reduce creep or lateral shifting, control water flow, etc. To best provide this function, stabilizing chemicals that develop a cementitious matrix are preferred. These additives include such materials as cement slurry, fly ash or pozzolanic earth in lime or cement slurries, sodium silicate (water glass) mixed with a strong acid, or methyl methacrylate polymerized by a peroxide catalyst.[17] The basic nature of the individual soil particle is unchanged; the particles are merely cemented together by the cementitious material filling the void (or pore) space.

As a rule, the stabilizing materials are introduced into the soil, through some variation of pressure injection, often to depths of 10 to 20 ft (3 to 6 m). Injection pipes are generally mechanically driven or washed down by water to total depth. Once grout injection commences, the pipe is slowly withdrawn. In other words, injection proceeds from the bottom toward the surface and generally continues at each level until either refusal or some predetermined volume is placed. Varying from one type of material to another usually involves little more than changing the mixer and/or pump. For more information on this subject, refer to Chap. 4.

7.6.1 Pressure Grouting

Pressure grouting is also often used to improve the bearing capacity of impermeable and permeable soils. The key is to either compress the soil material to a level above the anticipated load or create a soil matrix that has the desired bearing capacity. For example, if a compressible organic/inorganic fill could be grouted to an actual pressure of 100 lb/ft² (7 kg/cm²), the soil theoretically could support a load of 14,400 lb/ft² (70,000 kg/m²), not taking into account any safety factors. (Actual pressure implies the true compressive pressure on the soil matrix, and not gauge pressure at the surface.) (also see Chap. 4.)

7.7 STABILIZING IMPERMEABLE SOILS

For our discussion, the impermeable soils are generally cohesive with appreciable clay content. As a rule, the clay constituent is usually expansive. The problems generated by a clay are influenced directly by variations in available moisture, the result of which is either shrinking or swelling. This volatile nature causes serious concern regarding the design, construction, and stability of foundations.

For example, a typical Eagle Ford soil (Dallas, Texas) with a PI of about 42 exhibits a confined swell pressure of about 9000 lb/ft² (44,000 kg/m²) when the moisture content is increased from 23 to 26 percent. In this example, the problem clay constituent is montmorillonite, which is present at up to 50 percent of the total solids' volume.[102,103] Considering that the preponderant weight of a residential or light commercial structure is carried by the perimeter beam and that load is appreciably less than 1000 lb/ft³ (4880 kg/m²) for single-story construction, it is obvious that structural instability is imminent. If the soil's upward thrust is 9000 lb/ft² (44,000 kg/m²) and the maximum structural resistance is less than 1000 lb/ft²), the building will eventually rise. The interior floor area often represents loads as low as 50 to 100 lb/ft² (240 to 490 kg/m²).

Because of the difference in structural resistance as well as the heterogeneous nature of the soil, the uplift, or heave, is seldom, if ever, uniform. The secret is to deny the soil the 3 percent change in moisture or to alter the properties of the clay constituent to the extent that influences of differential water are neutralized. The former is discussed in Chap. 8; the latter can be accomplished by treating the problem soil with certain chemical agents. The stabilization procedure depends to some extent on one's comprehension of the nature of the specific clay constituents. When the subject soil is basically an expansive clay, compaction alone is most often inadequate to prepare the site for foundation support. It is desirable to alter the soil's behavior by either the use of chemicals, pressure grouting, or occasionally, perhaps a combination of both. Overconsolidation of shallow or surface soils should be avoided.

7.7.1 Chemical Stabilization

The soil upon which a foundation is supported influences or dictates the structural design and ultimate stability of the constructed facility. Earlier chapters have dealt with the pertinent physical properties inherent to and desirable within a bearing soil. Some discussion has been devoted to problem soil components such as the expansive clays. Criteria for overcoming the clay problems through design and maintenance of the foundation have been discussed. This section addresses other options: (1) impart beneficial properties to the otherwise problem soil or (2) alter the offending clay constituent to reduce or eliminate the volatile potential. Chemical stabilization represents a classic approach to this problem and can be separated into two categories: (1) permeable soils (which generally include noncohesive materials such as sand, gravel, organics, and occasionally silts), discussed in preceding paragraphs, and (2) the nonpermeable soils that generally are cohesive in nature and contain the clays (e.g., montmorillonite, attapulgite, chlorite, illite, and kaolinite). These are also referred to as expansive soils (except kaolinite) and often need to be stabilized to control shrink and swell.

Actually, *stabilization* can be a little misleading. Most chemicals are designed to abate swell. To eliminate settlement, transpiration would need to be eliminated, which translates to no live plant roots. This option is not looked on favorably by most homeowners. Most products offer selective side benefits such as increased permeability, increased bearing strength, and perhaps, some decrease in shrink. Still the principal intent is to reduce swell.

One of the less-expensive stabilization methods is to maintain the in situ soil moisture at a constant level. This can sometimes be easier said than done. Several approaches have been proposed in the literature, such as (1) sophisticated watering systems[15–17] and (2) moisture barriers.[15–17,26,79] Proper watering has shown the most potential. The other choice is to chemically modify the clay to control its volatile nature.

7.7.2 Inorganic Chemicals

Stabilization by inorganic chemicals has been widely used. The principal mechanism for this reaction is that of cation exchange. The increased positive charges hold the clay platelets closer together, inhibiting swell; also, cementatious benefits are sometimes realized.[15–17] Lime, Ca $(OH)_2$ has been used for over 30 years—particularly for highway construction. The lime is intimately mixed with the base and/or subbase soils (tilled into the matrix), watered down, and compacted. The calcium cation exchanges with clay constituents and may produce some pozzolanic reactions with silicas. Both actions afford stabilization to the soil.[15–16] Tilling has also been used successfully to some extent in new residential construction. However, the process has not been generally accepted for remedial applications. This is largely due to cost, problems in obtaining an adequate mix of the lime into the soil matrix, questionable results, and construction delay due to the wet site. A modification of the lime application was introduced in the 1960s when a lime slurry [Ca (OH)$_2$ in water] was pressure injected into the soil (LSPI). In a further effort to facilitate penetration of the slurry into the soil, a surfactant (organic) was frequently added. Pressure lime injection has experienced moderate success in new construction. The principal drawback still focuses on lime's very low solubility in water and the vertical impermeability of the expansive soils. This technique has been tried on remedial projects with much less success.[3,15–17,26,79] Figure 7.1 documents a monumental failure for LSPI. This depicts a foundation severally damaged from movement induced by pressure lime injection. Following the initial treatment, no appreciable amount of leveling was noted (as would be expected); however, it was assumed that stabilization had been accomplished. Complete cosmetic repairs were performed. Two years later the observations shown in Fig. 7.1 were noted. The photos speak for themselves. The elevation taken (Fig. 7.1c) suggests a "new" differential movement in the range of 5 in (12.5 cm). Note also the washboard nature of the slab surface. This problem was ultimately remedied by underpinning and mudjacking the slab foundation. In other applications the stabilizing chemical might involve a mixture of potassium chloride (KCl), an organic surface active agent (surfactant) and perhaps sulfuric acid (H$_2$SO$_4$). Aside from the obvious hazardous aspects of handling it, the reaction of sulfuric acid with lime, already present in the soil, offers yet another serious concern. Lime and sulfates react within the wet clay soils to produce ettringite—a calcium, aluminum, sulfate hydrate (3 CaO-Al$_2$O$_3$-3 CaSO$_4$-32 H$_2$O). This product has an unpredictable and uncontrollable swell.[3,82,85,106] (The use of lime in clay soils that contain sulphates has experienced the same problems.) The use of KCl plus a surfactant would probably prove to be a better, certainly safer, product without H$_2$SO$_4$.

7.7.3 Organic Chemicals

Over the years several organic-based products have been used to stabilize the clays, with varying degrees of success. One such product, Soil Sta, is the lone product for which reliable data can be found in the literature. Essentially, this product is a mixture of a polyquatenaryamine and surfactant in water. The product is neither toxic nor hazardous. In over 6000 applications, not one single failure to abate intolerable swell has been recorded. In most tests the soil swell potential was reduced to less than 1 percent at moisture contents of 17 percent or higher (PL 30 percent) (see Fig. 7.2). The product has been tested to increase the soil's permeability by a factor of 8, increase the soil resistance to shear twofold, and reduce soil shrinkage by ranges of 11 to 50 percent.[15–17] The organic chemicals have an extremely high solubility in water, acceptable permeation into soil matrix, reasonable cost, and a most effective performance.

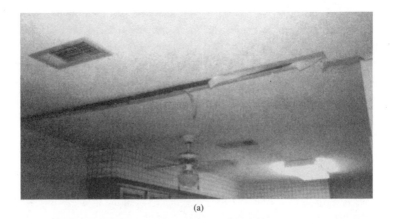

(a)

(b)

FIGURE 7.1 Failure due to pressure lime injection. (*a*) Sheetrock cracks in wall and ceiling; (*b*) separation in brick mortar with lateral displacement of brick.

One concern, however, is choosing the proper condition under which the use of the chemical(s) is cost effective. When the primary concern is to abate swell, the use of Soil Sta is probably not warranted if the soil's natural moisture content (W%) is in the range of the PL. At this moisture, a high percentage of the soil's swell potential has already been realized.[15–17] In other words, without the introduction of a chemical stabilizer, additional water is not likely to cause significant soil swell. On the other hand, if the soil's in situ moisture is appreciably lower than the PL, Soil Sta could be an effective choice. The greater the range between W% and PL, the more effective will be its performance. The product is most effective on the montmorillonite clays.

(c)

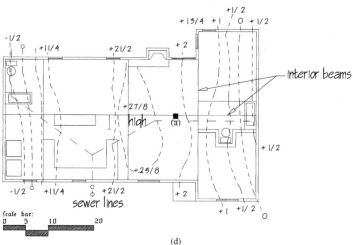

(d)

FIGURE 7.1 (*Continued*) Failure due to pressure lime injection. (*c*) Separation in brick frieze; (*d*) relative elevation survey.

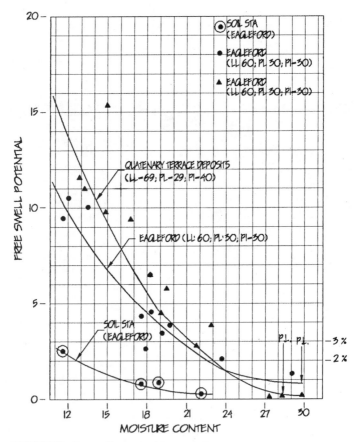

FIGURE 7.2   Free swell versus moisture content.

7.8   CLAY MINERALOGY

Basically, the surface clay minerals are composed of various hydrated oxides of silicon, aluminum, iron, and, to a lesser extent, potassium, sodium, calcium, and magnesium. Because clays are produced from the weathering of certain rocks, the particular origin determines the nature and properties of the clay. Chemical elements present in a clay are aligned or combined in a specific geometric pattern referred to as structural or crystalline lattice, generally sheetlike in appearance. This structure, coupled with ionic substitution, accounts principally for both the various clay classifications and their specific physical and chemical behaviors.

By virtue of a loose crystalline structure, most clays exhibit the properties of moisture absorption (and ion exchange). Among the more common clays, with the tendency to swell in decreasing order, are montmorillonite, illite, attapulgite, chlorite, and kaolinite. Figure I.1 shows the areas of general and local abundance of high-clay, expansive soils. The darker areas indicate those states suffering most seriously from expansive soil problems.

The data in the figure (provided by K. A. Godfrey, *Civil Engineering,* October 1978) indicate that nine states have extensive, highly active soils and eight others have sufficient distribution and content to be considered serious. An additional 10 to 12 states have problems that are generally viewed as scattered or relatively limited. As a rule, the 17 "problem" states have soil containing montmorillonite, which is, of course, the most expansive clay. The 10 to 12 states with so-called limited problems (represented by the lighter coloring) generally have soils which contain clays of lesser volatility such as illite and/or attapulgite or montmorillonite in lesser abundance.

A specific clay may adsorb and absorb water to varying degrees—from a single layer to six or more layers—depending on its structural lattice, presence of exchange ions, temperature, environment, and so on. The moisture absorbed may be described as one of three basic forms: interstitial or pore water, surface adsorbed water, or crystalline interlayer water. This combined moisture accounts for the differential movement (e.g., shrinking or swelling) problems encountered with soils. To control soil movement, each of these forms of moisture must be controlled and stabilized.

The first two forms, interstitial (or pore) water and surface adsorbed water, are generally accepted as capillary moisture. Both occur within the soil mass external to individual soil grains. The interstitial, or pore, water is held by interfacial tension and the surface adsorbed water, by molecular attraction between the clay particle and the dipolar water molecule. Variations in this combined moisture are believed to account for the principal volume change potential of the soil. Refer also to the Introduction.

Soils can take on or lose moisture. Within limits, this moisture exchange involves pore or capillary water, sometimes referred to as free water. (It is recognized that capillary water can be transferred by most clays to interlayer water, and vice versa. However, the interlayer water is normally more strongly held and, accordingly, most stable. This will be discussed at length in the following paragraphs.)

In a virgin soil, the moisture capacity is frequently at equilibrium even though the water content may be well below saturation. Any act that disturbs this equilibrium can cause gross changes in the moisture affinity of the clay and result in either swelling or shrinking. Construction, excavation, and/or unusual seasonal conditions are examples of acts that can alter this equilibrium. As a rule, environmental or normal seasonal changes in soil moisture content occur only very close to the ground surface. That being the case, it would appear that for on-grade construction, it should be sufficient to control the soil moisture only to this depth.

At this point it should be emphasized that, for capillary water to exist, the forces of interfacial tension and/or molecular attraction must be present. Without these forces the water would coalesce and flow, under the force of gravity, to the phreatic surface (top boundary of water table). The absence of these forces, if permanent, could fix the capillary moisture capacity of the soil and aid significantly in the control of soil movement. Control or elimination of soil moisture change is the basis for chemical soil stabilization.

Interlayer moisture is the water situated within the crystalline layers of the clay. The amount of this water that can be accommodated by a particular clay depends on three primary factors: the crystalline spacing, the chemical elements present in the clay crystalline structure, and the presence of exchange ions. As an example, bentonite (sodium montmorillonite) will swell approximately 13 times its original volume when saturated in fresh water. If the same clay is added to water that contains sodium chloride, the expansion is reduced to about threefold. If the bentonite clay is added to a calcium hydroxide solution, the expansion is suppressed even further, to less than twofold. This reduction in swelling is produced principally by ion exchange within the crystalline lattice of the clay. The sorbed sodium ions (Na^+) or calcium ions (Ca^{2+}) limit the space available to the water and cause the clay lattice to collapse and further decrease the water capacity.

As a rule (and as indicated by this example), the divalent ions such as Ca^{2+} produce a greater collapse of the lattice than the monovalent ions such as Na^+. An exception to the preceding rule may be found with the potassium ion (K^+) and the hydrogen ion (H^+). The potassium ion, because of its atomic size, is believed to fit almost exactly within the cavity in the oxygen layer. Consequently, the structural layers of the clay are held more closely and more firmly together. As a result, the (K^+) becomes abnormally difficult to replace by other exchange ions. The hydrogen ion, for the most part, behaves like a divalent or trivalent ion, probably through its relatively high bonding energy. It follows that, in most cases, the presence of H^+ interferes with the cation exchange capacity of most clays. This has been verified by several authorities. R. G. Orcutt et al. indicate that sorption of Ca^{2+} by halloysite clay is increased by a factor of 9 as the pH (OH^- concentration) is increased from 2 to 7.[81]

Although these data are limited and qualitative, they are sufficient to establish a trend. R. E. Grim indicates that this trend would be expected to continue to a pH range of 10 or higher.[45,46] {The citation exchange at high pH, particularly with Ca^{++}, holds significant practical importance, that is, the stabilization of expansive soils with lime [$Ca(OH)_2$].} The pH is defined as the available H^+ ion concentration. A low pH (below 7) indicates acidity, 7 is neutral, and above 7 is basic. Cement or lime stabilization of roadbeds represents one condition in which clays are subjected to Ca^{2+} at a high pH. It should be recognized that under any conditions the ion-exchange capacity of a clay decreases as the exchanged-ion concentration within the clay increases. Attendant on this, moisture-sorbtion capacity (swelling) decreases accordingly.

The foregoing discussion referred to changes in potential volumetric expansion brought about by induced cation exchange. In nature, various degrees of exchange preexist, giving rise to widely variant soil behavior even among soils containing the same type and amount of clay. For example, soils containing Na^+-substitute montmorillonite are more volatile (expansive) than are soils containing montmorillonite with equivalent substitution of Ca^{2-} or Fe^{3+} because Na^+ is more readily replaced by water and the absorption and/or adsorption of the water causes swell. Also, the high valence tends to hold the clay platelets in closer contact, inhibiting the intrusion of water.

To this point, the discussion has focused on inorganic cation exchange. However, published data indicate that organic ion adsorption might have even more practical importance to construction problems.[35] The exchange mechanism for organic ions is basically identical to that discussed above, the primary difference being that, in all probability, more organic sorption occurs on the surface of the clays than in the interlayers and, once attached, is more difficult to exchange. Gieseking[43] reports that montmorillonite clays lost or reduced their tendency to swell in water when treated with several selected organic cations. The surface adsorbed water (or double diffuse layer) that surrounds the clay platelets can be removed or reduced by certain organic chemicals. When this layer shrinks, the clay particles tend to pull closer together (flocculate) and create macropores or shrinkage cracks (intrinsic fractures). The effect of this is to increase the permeability of the expansive soil.[45]

This action should be helpful to reduce ponding, reduce run-off, and facilitate chemical penetration for stabilization. The extent of these benefits depends on the performance of the specific chemical product. Specific chemical qualities tend to help promote the stabilization of expansive clays. Among these are high pH, high $[OH]^-$ substitution, high molecular size, polarity, high valence (catonic), low ionic radius, and highly polar vehicle. Examples of organic chemicals that possess a combination of these features include polyvinylalcohols, polyglycolethers, polyamines, polyquarternryamines, polyacrylamides, pyridine, collidine, and certain salts of each. Because none of the above organic chemicals possess all the desired qualities, they are generally blended to enhance the overall performance. For example, the desired pH can be attained by addition of lime $[Ca(OH)^2]$, hydrochloric acid (HCl), or acetic acid (C_2H_3OOH); the polar vehicle is generally satisfied by dilution with water; high molecular size can be accomplished by polymerization; surfactants can be utilized to improve penetration of the chemical through the soil, and inorganic cations can be supplied to provide additional base exchange. (For more detailed information concerning the chemical reactions of base exchange in montmorillonite refer to Secs. 6.6.3 and 6.6.4 of *Foundation Behavior and Repair.*[16]) Generally, the organic chemicals can be formulated to be far superior to lime with respect to clay stabilization. Organic chemicals can be selected that are soluble in water for easy penetration into the soil. Chemical characteristics can be more finitely controlled and the stabilization process can be more nearly permanent. About the only advantages lime has over specific organics, at present, are lower treatment cost, more widespread usage (general knowledge), and greater availability.

The point will be made in later discussions that foundation repair generally is intended to raise the lower most areas of distressed structure to produce a more nearly level appearance. The repair could be expected to be permanent only if procedures were implemented to control soil moisture variations. This is true because nothing within the *usual* repair process will alter the existing conditions inherent in an expansive soil. Alternatively, chemical stabilization can alter the soil behavior by eliminating or controlling the expansive tendencies of the clay constituents when subjected to soil moisture variations. If this reaction is, in fact, achieved, the foundation will remain stable, even under adverse ambient conditions.

Several organic-based products are currently available to the industry. One such product, Soil Sta, is discussed in Sec. 7.6.5. This particular product was discussed principally because of the availability of the reliable data and its documented effectiveness.

7.8.1 Properties for a Clay Stabilizing Chemical

A superior chemical stabilizing agent must be both economical and effective. However, the definitions of these terms can be quite arbitrary. As a start, the economical aspect is assumed to be at a cost somewhat competitive with that for conventional lime. The effective aspect is more elusive. The chemical should be

1. Effective in reducing swell potential of clay.
2. Reasonably competitive with lime in cost, but more readily dispersed into the soil. (Lime is sparsely soluble in water and therefore difficult to use in expansive soils where cutting and tilling is inappropriate.)
3. Compatible with other beneficial soil properties.
4. Free from deleterious side effects such as a corrosive action on steel or copper, herbicidal tendencies, unpleasant smell, and hazards to health or the environment.
5. Easy to apply with few, if any, handling problems for the applicators or equipment.
6. Permanent.

At first, it might seem that the chemical should actually dehydrate clay, or in field terms "shrink the swollen soil." The problem occurs because such dehydration is most often unpredictable and nonuniform. It seems that a simpler approach would be to treat the clay to prevent any material change in the water content within the clay structure. Soil Sta was formulated to meet all the noted criteria. The mechanism by which this occurs is to both replace readily exchangeable hydrophilic cations (such as Na^+) and adsorb on the exposed cation exchange sites to repel invading water.

7.8.2 Chemistry of Cation Exchange

The chemistry of cation exchange and moisture capacity within a particular clay are neither exact nor predictable. A given clay under seemingly identical circumstances often has variations in cation exchange as well as the extent to which a particular cation is exchanged. The specifics of this exchange capacity dictate the water affinity and bonding to the clay structure. Refer to Sec. 5A, *Practical Foundation Engineering Handbook,*[17] for details on this topic.

In addition to the foregoing, many organic chemicals tend to shrink the double diffuse layer that surrounds the clay particles, causing the clay particles to flocculate and the soil skeleton to shrink. The net result is the formation of cracks (referred to as synersis cracks). The combination of these effects coupled with attendant desiccation increases the permeability of the clay.[38] The exposure of greater surface area may further facilitate the base exchange of certain organic molecules.[11,47,59] The chemical should then prevent swell upon reintroduction of water.

7.9 DEVELOPMENT OF A SOIL STABILIZING CHEMICAL

For some time, different groups have tested various chemicals as stabilizing agents to prevent the extensive swell of specific clays, in particular montmorillonite. These studies have involved the petroleum and construction industries. As a result, several materials exhibiting varying potential have evolved. However, the general emphasis has been on new construction. Considering all factors, hydrated lime [$Ca(OH)_2$], has been difficult to displace in types of applications in which a controlled, intimate mix with the clay/soil was feasible.

In remedial applications, such as stabilizing the soil beneath an existing structure, lime has its inherent shortcomings: It is difficult to introduce into the soil matrix with any degree of uniformity, penetration, and saturation. This is largely because (1) lime is sparsely soluble in water, and (2) the clay/soil needing stabilization is both impermeable and heterogeneous. A comprehensive state-of-the-art report on lime stabilization can be found in Dr. J. R. Blacklock's publications, the latest of which is referenced.[5]

However, the use of hydrated lime to stabilize montmorillonite clays can also create detrimental side reactions. In a study presented by Berry Grubbe,[48] lime stabilization of naturally expansive soils (Austin Chalk and Eagle Ford) resulted in significant heave to a pavement section. The heave reportedly was caused by the chemical reaction between lime and sulfates in the soil to produce ettringite (see Sec. 7.7.2). Obviously, although this report addresses distress to pavements, the soil swell (heave) problem would be the same in other applications of lime stabilization. Further, Tom Petry and D. N. Little describe soil heave brought about by sulfate introduction into clay soils containing lime.[85] In large part because of the foregoing, the successful use of lime in soil stabilization has not been documented for remedial applications.

As early as 1965, certain other surface-active organic chemicals were evaluated and utilized with some degree of success. One chemical utilized in the late 1960s and early 1970s was unquestionably successful in stabilizing the swell potential of montmorillonite clay. The chemical was relatively inexpensive and easily introduced into the soil. However, it maintained a "nearly permanent" offensive aroma that chemists were never able to mask. Generally, this product was a halide salt of the pyridine-collidene-pyrillidene family.

In the late 1970s, the quest began to focus more on the potential use of polyamines, polyethanol glycolethers, polyacrylamides, etc., generally blended and containing surface-active agents to enhance soil penetration.[104] It was found that certain combinations of chemicals seemed to be synergistic in behavior (the combined product produced superior results to those noted for any of its constituents). In the mid 1980s one such product was Soil Sta.

Note: The product Soil Sta is proprietary to the author. This presentation is not intended to be commercial. In fact Soil Sta is not marketed. Necessity suggests the focus on this particular product because similar laboratory and field data are not publicly available for any other stabilizer, except perhaps lime. Organic stabilizers function differently from lime, and no standardized testing procedures existed for the evaluation of these type products. The following discussions and data should prove beneficial to the others wishing to evaluate an organic chemical clay stabilizer. All descriptive data and information was supplied through the courtesy of Brown Foundation Repair and Consulting, Inc.; Dr. Cecil Smith, Professor of Civil Engineering, Southern Methodist University; Dr. Tom Petry, Professor of Civil Engineering, University of Texas, Arlington; and Dr. Malcom Reeves Soil Survey of England and Wales.

Soil Sta is basically a mixture of surfactant, buffer, inorganic cation source, and polyquaternaryamine in a polar vehicle. By virtue of its chemical nature, Soil Sta would be expected to have a lesser influence on kaolinite or illite than on the more expansive clays such as montmorillonite. Prior research has also indicated that soils exhibiting an LL less than 35 or a PI less than 23 (montmorillonite, content less than about 10 percent by weight) would not swell appreciably.[6] Hence, the soils utilized in the laboratory tests and field applications contained montmorillonite as a soil constituent above 10 percent.

Soil Sta was first subjected to laboratory evaluation in 1982 and field testing commenced in mid-1983. The laboratory tests indicated that Soil Sta

1. Reduced the free swell potential of montmorillonite clay (Fig. 7.2)
2. Appeared stable to repeated weather cycles (a simulated period of 50 years)
3. Increased shear strengths in some soils by two-fold
4. Increased soil permeabilities up to 40-fold
5. Reduced soil shrinkage by amounts varying from 11 to 50 percent[17,70,84]

By 1991 Soil Sta had been subjected to literally thousands of field applications with few, if any, failures. That is, less than 1 percent of the foundations treated with the chemical experienced recurrent movement. For those that did, there was a serious question as to the cause.

7.9.1 Pressure Injection

In special cases in the United States and for most applications within the United Kingdom, chemical injection is performed through a specially designed system (Fig. 7.3). In the system utilized, the Soil Sta is injected under pressure to some depth—usually 4 to 6 ft (1.2 to 1.8 m). Penetration of the stem is accomplished by pumping through the core and literally washing the tool down. It is difficult, if not impossible, to wash the stem down when the base course is rubble or coarse gravel. In this instance, a pilot hole through the fill material is required. This can be accomplished by using a paving breaker and steel point to penetrate the problem base. This done, normal placement of the stem can continue. Once positioned, the hand valves are switched to close the core and divert flow into the annular space and out the injection ports (Fig. 7.4). The stem can be raised during pumping to cover the desired soil matrix section. Generally, the treatment volume is determined on the basis of $^1/_8$ gal/ft^2 (0.5 mL/cm^2). In some instances, a particular soil might tend to resist Soil Sta penetration. Both the rate of penetration and the volume of chemical placed can be enhanced by utilizing hydraulic pulsation (high pressure of short duration) during the injection phase. Alternatively, the stem can be equipped with a packer assembly to selectively isolate zones (Fig. 7.5). This equipment permits high injection pressures and also allows zone selectivity.

From a practical viewpoint, minimal concern should be given to the *exact* volume of chemical injected into a specific hole. The primary intent is to distribute the treatment volume reasonably uniformly over the area to be treated. Time (days, weeks, or months, depending on the specific site conditions) will produce a nearly equal distribution.

A similar analysis would be true for depth of injection. Within a shallow depth [i.e., 6 ft (1.8 m) or less] the chemical will penetrate the same approximate depth and interval almost independently of the position of the stinger. Shallow penetration is accompanied by problems to confine the permeation of the

FIGURE 7.3 Chemical injection through a specially designed stem.

chemical into the matrix. Two changes could facilitate better chemical control: (1) The depth of penetration is increased substantially, that is, 15 to 20 ft (4.5 to 6 m), (2) a pulsing injection technique is used, and/or (3) packers or other positive seal methods are used to isolate each zone to be injected. Even in the latter case, true zone penetration may not occur if the placement pressure or specific soil characteristics favor communication between zones. To illustrate the point, no matter what precautions might be taken, the normal heterogeneous and fractured nature of the soil would tend to preclude *exact* placement of any specified volume.

At a pressure differential of about 3.5 lb/in² (24 kPa), the system illustrated in Fig. 7.4 would theoretically place about 12 gal/min (45 L/min), neglecting line friction. Hence, 10 s would be required to place 2 gal (7.6 L) of chemical. [This volume equates to ⅛ gal/ft² on a 4-ft (1.2-m) spacing pattern.] By timing the injection period and maintaining a reasonably constant supply pressure, an acceptably uniform treatment spread would result. Carelessness in either timing

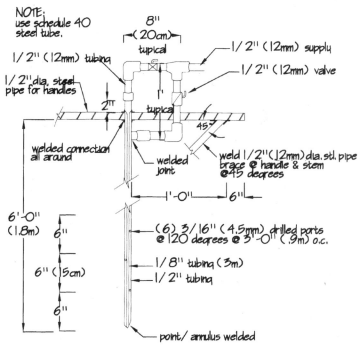

FIGURE 7.4   Pressure injection stem.

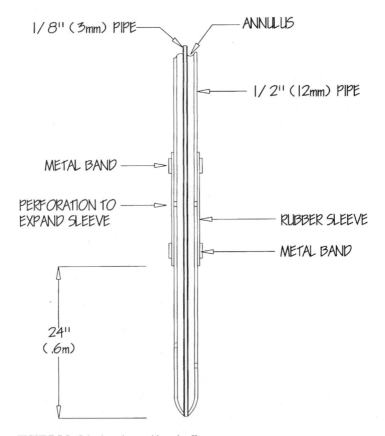

FIGURE 7.5   Injection stinger with pack-off.

or pressure would not be disastrous, so long as it was not blatant. [In field practice, a pressure differential of about 60 lb/in² (414 kPa) delivered 2 gal (7.6 L) of chemical in 30 s through the stinger and approximately 60 linear feet (18.3 m) of ¹/₂-in (1.27-cm) I-D hose.] The following equations can be used to estimate velocities and pressure differentials.

The annular velocity is

$$V_a = \frac{Q}{A} = 1.84Q \text{ ft/s} \tag{7.1}$$

where annular area $A = 0.175$ in² and Q is in gallons per minute. The port velocity is

$$V_P = \frac{4Q}{\pi n C_D} \quad \text{ft/s} \tag{7.2}$$

Where $D_D = {}^3/_{16}$ in, the number of ports $n = 6$, and Q is in gallons per minute.

Equation 7.2 reduces to

$$V_P = \frac{Q}{\quad}$$

where $A_p = D_p^2 \, \pi n/4$. The orifice discharge coefficient C_D can be assumed to be 0.8.[20] The pressure differential is

$$P = \rho_f \frac{(V_P^2 - V_a^2)}{149} \quad \text{lb/in}^2 \tag{7.3}$$

where ρ_f is the specific gravity of the fluid (water = 1.0). The force developed from hydraulic pressure is

$$F = PA \tag{7.4}$$

where F = force, lb_f or kg; P = pressure, lb/in^2 or kg/cm^2; and A = area, in^2 or cm^2.

Equation 7.4 is used to illustrate the factors creating a lifting force. Generally, pressure injection of chemical soil stabilizers does not involve lifting, as mudjacking or pressure grouting would. In fact, as a rule, chemical injection pressure could then be estimated from the rearrangement of Eq. 7.4 as follows:

$$P = \frac{F}{A}$$

7.10 WATER BARRIER

The use of chemical soil stabilizers (Soil Sta in particular) has focused renewed interest on the use of water barriers.[18] In areas where the in situ soil moisture is relatively high (nearing or exceeding the soil's plastic limit), a combined technique utilizing Soil Sta to control soil swell and a moisture barrier to prevent soil moisture loss (shrinkage) appears to have some merit. The use of Soil Sta in a relatively wet soil was predicated on the possibility that at some future point the soil moisture might be reduced substantially for a period of time and then increased back to or near the original moisture. This approach is being considered in the United Kingdom. In this instance, the slit trench [approximately 3 in (7.6 cm) wide by 60 (1.5 m) deep, typically] is dug as near to the foundation perimeter as conditions permit and filled with concrete. This creates the moisture barrier (see also Fig. 9.5 and Sec. 9.5.2). Soil Sta is injected according to the selected procedure described in preceding paragraphs. The barrier is intended to prevent the peripheral loss of soil moisture due to either evaporation or transpiration. (Soil Sta has minimal effect on soil moisture loss to transpiration. Otherwise, the chemical would be detrimental to vegetation.) Obviously, the foregoing procedure is designed and intended to stabilize the soil moisture (and foundation) "as is" with negligible, if any, leveling. Where leveling is desired, or necessary, conventional methods are employed. They may or may not include this stabilization technique (see also Secs. 9.10.9 and 10.2.2).

7.11   IRRIGATION

An irrigation system similar to that depicted in Figs. 9.1 and 9.2 could also overcome any peripheral loss of moisture. This system simply replaces any moisture otherwise lost from the soil by either evaporation or transpiration. The key to the effectiveness of this approach lies principally within the metering and monitoring equipment. The moisture returned by the system should be carefully controlled to replace water lost but at the same time maintain a constant soil moisture. Special care should be exercised not to provide an overabundance of water. This oversight could (and often does) result in the most serious problem of soil swell and foundation upheaval.

7.12   COST

Computing the cost for reliable, widespread, chemical stabilization is very difficult. This is generally because

1. Many products are proprietary, and application procedures vary.
2. There is little history or cost data in the publications.
3. Treatment specifications and applications vary broadly.
4. There is no basic standard for acceptable performance. Standard Atteberg limit tests offer little value.

In fact, the only cost figures, which the author will stand behind, are those for the chemical Soil Sta. Other data was acquired secondhand, generally, by word of mouth. Again, the labor costs used in placement costs should be computed as a relative rate (based on unskilled labor) at $6/h in 1998 U.S. dollars.

Method	Cost
Lime stabilization	
Mechanical mixing	$0.27/ft^2 6 in (30 cm), lift with 6 percent lime
Pressure injection	$0.23/ft^2 to 7 ft (2.1 m)
	$0.15/ft^2 to 4 ft (1.2 m)
	$0.13/ft^2 large area
Chemical stabilization (pressure injection)	
Chemical A	$2.17/ft^2 to 6 ft
Chemical B	$0.20/ft^2 to 6 ft

The following figures are for chemical stabilization using Soil Sta:

Area	New construction, $	Remedial, $
1000 ft^2	1.30	
2000 ft^2	0.65	0.65
4000 to 8000 ft^2	0.50	0.65
8000+ft^2	0.41	0.60

Lime stabilization accomplished by mechanical mixing is normally bid on the basis of dollars per square yard. This cost was changed to dollars per square foot in an effort to present a better view of the comparison. The prices for Soil Sta include

1. The use of $^1/_8$ gal of chemical per square foot.
2. Sufficient chemical to treat the soil to a depth of 6 ft (1.8 m).
3. Injection sites 5 ft (1.5 m) OC to a depth of 4 to 5 ft (1.2 to 1.5 m) for new construction. Injection holes on remedial applications are fewer in number (wider spaced) due to the specifics of the job.
4. Includes a 5-ft (1.5-m) apron around the footprint of the foundation.

Only Soil Sta offered numbers specifically for remedial applications.

CHAPTER 8
CAUSES OF FOUNDATION PROBLEMS

8.1 INTRODUCTION

Serious consideration was given regarding the location of this chapter within the text of the book. Should it appear in front of the "repair" discussion? Would the reader be better served if the cure for the problem was presented prior to the cause? The latter was finally selected. The major cause for foundation failures on expansive soil is water—too little or too much. Expansive soils suffer much more from this than nonexpansive soils. However, both can be affected. The variation of water results in settlement (expansive or nonexpansive) and upheaval (expansive). Factors other than soil moisture variations are discussed in Sec. 8.4.

8.2 SOIL MOISTURE LOSS

Soil moisture loss occurs principally from evaporation, transpiration, or a combination of both (evapotranspiration). Figure 8.1 offers a drawing typical of settlement and a photograph of an actual occurrence.

8.2.1 Evaporation

Evaporation represents the natural loss of soil moisture to the elements, heat and wind. In expansive (cohesive) soils the rate of loss is fairly slow and the depth of loss somewhat limited, both the result of very low permeabilities.

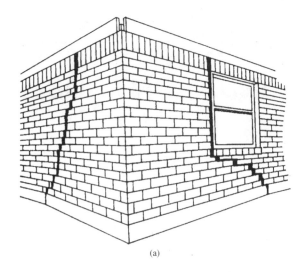

(a)

(b)

FIGURE 8.1 (*a*) Typical foundation settlement. (*b*) Foundation settlement at the corner of the perimeter beam caused the very noticeable separation in the brick mortar. In this instance, the crack is in excess of 2 in (5 cm). Note that the brick has also slipped.

The depths to which the moisture loss occurs is referred to as the soil active zone (SAZ). Most published authorities suggest that the SAZ may extend to depths as great as 3.2 to 12.8 ft (1 to 4 m).[16,17,42,53,62,68,97,103] However, a very high percentage of the *total* moisture loss (evapotranspiration) occurs at reasonably shallow depths. For example,

1. A study performed in Dallas, Texas, suggests that 87 percent of the total soil moisture loss occurs at depths above 3 ft (0.9 m).[103]

2. Holland et al. published data on an Australian soil that described an effective depth of 4 ft (1.25 m).[42]

3. Another study of several diverse areas within the United States indicates that over 80 percent of the total soil moisture loss occurs within the top 4.5 ft (1.5 m).[68]

4. A similar study of an Israeli soil suggests that 71 percent of the total soil moisture loss occurs at depths above 3.2 ft (1 m).[62]

5. A study by Sowa presented data that suggests an active depth in a Canadian soil of 1 to 3.2 ft (0.3 to 1.0 m).[97]

6. An English study describes a London soil where the range of principal moisture loss is 3 to 3.5 ft (0.9 to 1.1 m).[54]

Evaporation losses in expansive soils account for the very noticeable cracks in the earth. As the soil becomes drier, the cracks grow wider and deeper until the soil moisture content adjacent to the crack might approach the shrinkage limit (SL).[16,65,99] The soil remains desiccated until water is once again made available. The soil then absorbs water and swells. The soil moisture content might then rebound to exceed the PL.[7,16,65,69,99] Overall soil (and foundation) movements are principally active when their in situ or "natural" moisture content is between the SL and PL. However, shrinkage potential continues at moisture content above the PL. In fact, the upper limit for soil shrinkage (evapotranspiration) probably extends to the field capacity, a point somewhere between the LL and PL.[69] Figure 8.2 depicts the relationship between soil suction and volume. The range of soil moisture loss, due to *transpiration,* which would be expected to influence *foundations,* is between the field capacity and the plant wilt. Plant wilt is generally located at some point between the PL and the SL.[26,69] As soil suction increases, volume change decreases. (Soil suction describes the total moisture migration due to the combined forces of osmosis and capillary action. Thus soil suction is a measure of the soil's capacity for water.) This relationship might lead one to assume that a foundation distressed by soil settlement would be *completely* restored by replenishing the lost water. This is not normally the case. Historically, as the moisture content cycles, the foundation moves up and down. However, each wet cycle leaves the foundation somewhat short of its original grade.[69] Figure 8.3 shows shrink-swell profiles versus moisture content. In this study the dry

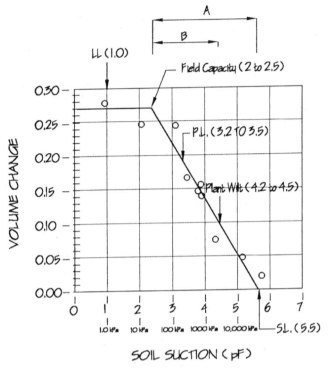

FIGURE 8.2 Range of relative volume change. A=evaporation and transpiration; B=transpiration. After McKeen, 1992.[169]

density was kept constant at 107.0±0.6 pcf (1714±9.6 kg/m³) and initial moisture content (W%) varied from 15.1 to 22.3 percent. This range of moisture essentially covers the spectrum between the SL and PL for the particular soil and is believed to be within the critical moisture range defined by Popescu.[69] A surcharge pressure of 1 psi (6.9 kPa) was applied at the initial moisture content (W%). At moisture contents of approximately 16 percent (SL) and 22 percent (PL), shrinkage is equivalent to swell. Between these points (A and B), shrinkage exceeds swell. At a moisture content below point A (SL), shrink is negligible, and swell increases rapidly. Moisture contents greater than the PL (B) depict a rapid decline in both shrink and swell. Only at points A and B are the two equal at the same moisture content. Hence, watering a foundation, once it has settled, might arrest further movement but is not likely to bring it back to original level.

8.2.2 Transpiration

Transpiration is arguably a bigger thief of soil moisture than is evaporation (see also Chap. 9). The removal of soil moisture attributed to transpiration declines substantially during the plant dormant season—perhaps to less than 10 percent of the requirement during growth season. The depth of primary soil moisture loss due to transpiration is generally limited to the top shallow soils, 1.0 to 2.0 ft (0.3 to 0.6 m).[16,17,49,50,54,98,99] Under normal conditions the shallow "feeder" roots account for probably 90 percent of the

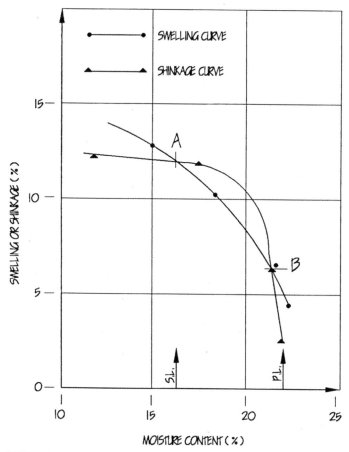

FIGURE 8.3 Effects of moisture content on swelling and shrinkage. After Chen, 1988.[26]

plants' total requirement for nutrients and water. [In arid climates, deeper roots (i.e., tap root) produce much of the water requirements for the plant. However, deep-seated moisture losses do not generally influence the stability of lightly loaded foundations.] The lateral extent over which the shallow roots are active is generally limited to the canopy area of the plant or tree.[15–17,42] As a matter of curiosity, what would be a reasonable estimate for the soil moisture loss due to transpiration? This question has many answers because transpiration depends largely upon the type, size, and density of vegetation as well as the ambient wind and temperature. Nonetheless, it is possible and practical to use reasonable assumptions and arrive at a representative number. Dr. Don Smith, Botany Professor, University of North Texas, suggested a "reasonable" value of water input for a 10 in (25 cm) diameter. Quercus Stellata (Post Oak) would be about 50 gal/day in a semitropical climate. Assuming a 30-ft (9-m) canopy (root spread), the moisture loss per surface square foot would be 0.07 gal (2.34 L), or 16.3 in^3 per day. A normal lawn watering program for the same geographic area would distribute 144 in^3/ft^2 (0.62 gal/ft^2 or 18.6 l/m^2) twice weekly. Over a 1-week period, transpiration by the tree would account for 114 in^3/ft^2 and the watering would add 288 in^3/ft^2 (0.093 m^2). The surplus water 174 in^3 would go into dead storage within the soil until removed by evaporation or transpiration. The main point is that tree roots would have adequate moisture and would not seek the inhospitable soil conditions beneath the foundation. This analogy excludes both the effects other vegetation might have on overall transpiration and the natural sources for precipitation (dew, rainfall, etc.). Generally speaking, if the vegetation has a healthy green color, the root moisture need is being satisfied.

The biggest threat that roots pose to foundations might be the presence of surface roots beneath a shallow foundation beam. As the root grows, a shallow beam can be literally "jacked" upward, or heaved. Tree roots do not cause the foundation problems for which they are blamed.[16,26] Pier and beam foundations have interior floors isolated from the bearing soil, and slab foundations have perimeter beams that normally extend well into or below the surface root zone.

Removing a *mature* tree in close proximity to a foundation can cause more problems than are cured. The decaying plant roots can produce both permeable channels that might funnel water beneath the slab foundation, resulting in upheaval, and voids, which result in settlement. On the other hand, tree roots can enhance foundation stability by increasing the soils resistance to shear.[28,107] Over the first few years one foundation repair contractor requested his drilling and excavation superintendents to keep track of instances where tree roots were encountered *beneath* the perimeter beam. At this point no effort was made to specifically identify the type of root. Data was also not recorded concerning type, age, size, or proximity of tree to the perimeter beam. The superintendents were instructed to merely record roots and type of foundation. On the average a minimum of

20 piers or spreadfootings were installed per day. Assuming only a 5-day work week, this would provide about 100 weekly observations, or 5200 per year. In cases where the depth of the beam below grade exceeded 18 in (0.45 m) roots were reported in less than 0.5 percent. When the perimeter beam depth was approximately 12 in (0.3 m), the instance increased to about 1 percent. When beam depths were less than about 6 in (0.15 m), the incidence increased to about 4 percent. Clearly, the data do not suggest a real-world reason to panic or, for that matter, even have a particular concern. The quality of these data might be somewhat questionable because the controls and specifics were less than "scientific." This is an excellent area for more research.

8.2.2.1 Summary: Tree Roots versus Foundation Distress: Slab Foundations on Expansive Soil. Botanists, horticulturists, and agronomists uniformly seem to agree that

1. The canopy of a tree determines the aerial extent of the feeder root system. The roots will extend only a short distance beyond the drip line of the canopy. According to Don Smith Ph.D., University of North Texas, a *reasonable* number to describe the radial root pattern would be $1.25C$, where C is the *radius* of the canopy, or the radial distance from the trunk to the drip line. (The 1.25 factor is conservative. For most trees under most conditions the factor is likely closer to 1.1 or less.)

2. Feeder roots tend to be quite shallow—no deeper than 1 to 2 ft (0.3 to 0.6 m).[49,50,98] This preference is promoted by the fact that tree roots prefer rich, aerated, and noncompacted soil.[50] Deeper roots may also contribute to the nutritional needs of the tree dependent upon the species and size of the tree and availability of water and to a lesser extent on ambient temperatures. (In conditions of drought the deeper roots play an increased roll. Shallow roots can become active when shallow moisture is restored.)

3. Soil moisture loss due to transpiration has been determined to extend to depths below the soil water belt.[16] However, in each of these cases, the *preponderant* loss (80 to 90 percent) occurs at relatively shallow depths, generally within the top 3 ft (1 m).[53,62,68,97,102,103,] T. J. Freeman et al. suggest that soil moisture loss below 6.6 ft (2 m) may not materially influence foundation behavior.[42]

4. Chen states "the end result of shrinkage around or beneath a covered area seldom causes structural damage and therefore is not an important concern to soil engineers."[26] This analysis obviously introduces the presence of the foundation slab into the soil moisture loss equation.

5. The presence of shallow roots can be beneficial to foundation stability because their presence increases the soil's resistance to shear.[28,107]

6. McKeen suggests that the range for relative volume change exists within the span between the PL and SL.[26,69] Classically this moisture loss tends

to involve pore water, the loss of which does not necessarily relate to soil shrinkage.[26,28,69] [Water bonded to or within the clay particles can be transferred to pore water, but temperatures above 212°F (100°C) are required.]

7. One inch [2.5 cm of water spread over 1000 ft^2 (93 m^2)] equates to 623 gal (2360 L) of water. A tree with a canopy *area* of 1000 ft^2 ($C=D/2=36/2=18$ ft) might require 50 to 100 gal (379 L) of water per day to remain healthy.[1] Effectively, a little over 1 in (2.5 cm) of water, once a week, (whether supplied by watering or mother nature) would satisfy the needs of the tree, neglecting serious run-off. Roots would not be "encouraged" to encounter the hostile environment found beneath the foundation.

The foregoing lists facts developed by the referenced academicians, engineers, and geotechs. Some of their work did not involve "real-world" situations. That is, the conclusions were drawn from tests (1) made on exposed soil, (2) using slabs poured directly on the ground surface, or (3) wherein no reasonable effort was made to exclude evaporation from transportation. Sources supplying actual field data taken from situations involving real foundations suggest that the true root problem is really not much of an issue.[16] Data collected since 1964 and involving over 20,000 actual repairs, supply the following:

1. During the process of underpinning ordinary slab foundations (perimeter beam depth in excess of about 10 in (25 cm), roots were found beneath the beams in less than 2 percent of the excavations. (The average number of excavations per job was 13. This extrapolates to well over 250,000 observations.) In many of these instances, the roots noted may have been in place prior to the construction of the foundation.

2. Virtually never does the company supplying these data recommend either removal of trees or the installation of root barriers. Removal of a tree can create far more serious concerns than leaving it be. None of these projects required *later* removal of trees, which certainly suggests that roots were not a problem.[16]

3. If tree roots were truly a serious problem, why wouldn't all foundations of similar designs placed on similar soil and surrounded by similar trees experience problems?[16] Chapter 10 and Fig. 10.4 present a discussion on the impact of tree roots to foundation stability. In all, this slab foundation [a 30-in (76-m) perimeter beam] has eight pin oak trees growing along the west and south walls. The trees average 18 in (46 cm) in diameter with a height of 36 ft (11 m) and canopy diameter of 32 ft (9.8 m). The trees are within 6 ft (1.8 m) of the perimeter beam. The foundation shows no ill effects after 17 years. The soil is a fairly alluvial soil with a PI of about 42.

It would appear that many investigators confuse perimeter settlement with center heave. From hundreds of engineering reports on file it seems that the investigations arbitrarily assume the highest point within the foundation to be the bench or reference point. This obviously forces all surrounding measurements to be negative. Negative readings are generally associated with settlement. The preponderant conclusion then points to settlement. In lieu of another "culprit," nearby trees become the guilty party. This practice may be the result of either intent or oversight.

As a matter of fact, settlement is rarely the cause for foundation repair.[16,26] The preponderant cause for repair is upheaval, brought about by water accumulation beneath the slab. The source for the water can be natural (poor drainage) or domestic (utility/sewer leaks). The latter accounts for perhaps 70 percent of all foundation repair.

8.2.3 Evapotranspiration

Evapotranspiration is the combined loss of moisture due to both evaporation and transpiration. This is the mechanism that in the end accounts for soil shrinkage and foundation settlement. However, (to quote F. Chen) "problems relative to 'normal' settlement are comparatively few and minor. The end result of shrinkage around or beneath a covered area seldom causes structural damage and therefore is not an important concern to engineers."[26] Other publications have supported Dr. Chen.[15-17]

8.2.4 Remedy

Restoration of a foundation distressed by settlement is fairly straightforward. One needs merely to raise the settled area back to the as-built position. With slab foundations this involves mudjacking and, in some cases, underpinning with pier and beam foundations. The lower perimeter areas are underpinned and the interior floors are reshimmed. Refer to Chaps. 2, 3, and 5.

8.2.4.1 Recourse. If you suspect a serious problem, the best course of action is probably to consult an attorney. This does not in any form recommend or suggest litigation. Reliable advice is quite valuable. Recourse for settlement problems must generally involve negligent behavior on someone's part. For "new" construction, this could involve builders, engineers, architects, inspectors, contractors, and/or developers. The statute of limitations in many states is 10 years. Examples of negligent acts include (1) failure to properly design the foundation to accommodate the site conditions, (2) failure to comply with the plans or specifications, (3) negligent inspections during critical states of construction, (4) not exercising a prudent attempt to grade the lot to control drainage, or (5) a nonnegligence recourse could be insurers who offer structural coverage.

In preowned properties recourse again is generally limited to some form of negligence and could involve realtors, engineers, inspectors, or neighbors.

Events that might be covered here are (1) negligence of inspectors and/or engineers whose responsibility was to inspect the property to ensure that no "structural defects" existed, (2) realtors and sellers can be at fault under "latent defects" provisions, and (3) neighbors can be negligent if they create conditions that damage your property.

8.3 TYPICAL CAUSES FOR SOIL MOISTURE INCREASE (UPHEAVAL)

The foregoing paragraphs have related broadly to soil moisture gain or loss with no consideration being given to the source or cause for this moisture variation. Because upheaval is, by far, the most serious foundation concern—especially when dealing with concrete slab-on-grade foundations—this condition will be addressed first. Figure 8.4 is a drawing typifying upheaval and an actual photograph of interior slab heaval. The drawing is an oversimplification of upheaval but serves as an example. Note the preponderant vertical cracks, wider at the top; the separation of the frieze boards from the brick soldier course; the open cornice trim; and the bow in the slab. In the real world, the perimeter brick may not show the major results of upheaval because of the weaker interior slab. Often the exterior mortar joints are mostly level. The photograph shows upheaval as it normally appears. The central slab is clearly "domed." The downspout looks level, yet the ends are several inches off the floor.

Sources for soil moisture could include such diverse origins as

1. Rainfall
2. Domestic sources, which include watering or leaks in domestic supply or waste systems
3. Redistribution of soil moisture from wetter to drier soils
4. Subsurface water

8.3.1 Rainfall

Rainfall is perhaps the principal source for bulk water. However, as far as foundations are concerned, this threat is minimized due to the mere presence of the foundation itself. First, the structure isolates the foundation-bearing soil from contact with the rain. (The perimeter beam tends to restrict any lateral transfer of water.) Second, proper surface drainage adjacent to the beam forces surface water to flow away from the structure rather than be attracted to the soil beneath the foundation (run-off).

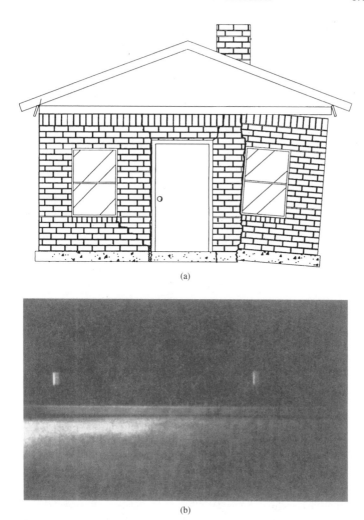

(a)

(b)

FIGURE 8.4 (*a*) A drawing typifying upheaval; (*b*) an actual photograph of interior slab heaval. In both examples, the "hump" in the floor is obvious.

8.3.2 Domestic Water

Domestic water is statistically the greatest concern, not so much from watering as from some form of leak. Watering can be controlled and, with the maintenance of adequate and proper drainage, any excess water can be safely directed away from the foundation. Leaks, on the other hand, are another matter. Supply line leaks can contribute a large quantity of water over a short period of time. However, a pressure leak is usually heard or noticed on the water bill and quickly repaired. The substantial threat, particularly to slab foundation, is the silent, sewer leak.[15–17,40,52,64,86] This problem can exist for years before detection. In fact, the existence of a leak is most often suspected or verified as a *result* of foundation damage. Upheaval can occur and advance rapidly with a rather extensive range and scope of damage. Very little water can cause serious distress. Under certain conditions, as little as six drops of water per hour, persistent over 12 months, can heave a slab foundation over 1 in (2.54 cm)[15–17,102] (see Chap. 1). Actual data indicate that over 70 percent of all slab foundation repairs are caused by upheaval (water accumulation beneath the slab). A high percentage of the recorded cases of upheaval have been linked to sewer leaks.[15–17]

A small cadre of engineers tend to deny the propensity and extent of the sewer leak problems.[15–17,26] The principal reason for this minority position might be traced to their clientele [i.e., insurance companies who have no liability for foundation losses attributable to causes other than "the accidental discharge of water from utilities" (or upheaval); settlement is not an insured act]. Refer to Texas Court of Appeals Case No. 13-92-C, Corpus Christi, December 16, 1993, Nicolau vs. State Farms Lloyds. Familiarity with this case is a must for those involved in foundation repair, appraisal, or engineering evaluation. (At the time of this writing the Appealant Court Decision has been appealed to the Supreme Court.)

Along these lines refer to Fig. 8.5. This drawing, prepared by one insurance company's engineer, depicts floor elevations taken inside a distressed slab foundation. The peak point is indicated to be about 6 in (15 cm) as shown. The basic central high is shown as 4.5 to 5.5 in (11 to 14 cm). The engineer acknowledges the two sewer leaks but states "since the leaks are not located near the heaving in the foundation, and there is no change in floor slope around the leak areas, I conclude that the reported underground plumbing leakage did not adversely affect the foundation." Review the engineer elevations: (1) The maximum grade differential is 6 in (15 cm). This heave would require a significant quantity of water. (2) The sewer line and leak are included within the central high ±5 in (13 cm). (3) Water following either the sewer ditch and/or the interior beams could very well account for the accumulation at the central point as well as the peak. (Some geotechs believe that water can flow through fill sand or coarse base materials.) (4) The foundation area between the cross beams (including bath leak) is aligned in the 5.0- to 5.5-in (13- to 14-cm) high spot. (5) The redistribution of *existing* soil moisture could not reasonably account for the magnitude and location of the heave.[16,26,102,103] The 6-in heave would require a seemingly impossible soil suction and "natural" water availability. The only logical source would be the sewer leak. (The location of both the sewer lines and the probable interior beams was added by the author based on "best available information.") The author had not inspected the property at the time of this analysis. If the issue goes to litigation, the author's inspection will become necessary.

Another curious point lies in the facts that (1) the group of engineers (who basically state that sewer leaks are not a significant cause of foundation distress) invariably recommend that the leaks be repaired, and (2) once the leaks are repaired, foundation movement generally ceases after some relatively short period of time. (The exception is most often the result of another *undetected* leak.) Both facts clearly suggest that the sewer leak is, in fact, the source of the problem.

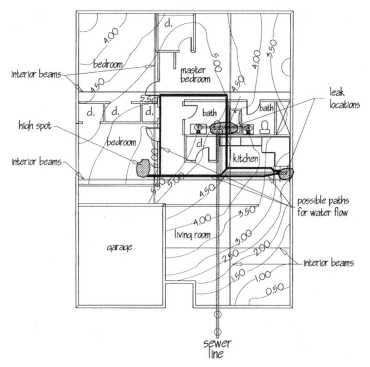

FIGURE 8.5   Heave versus floor elevation.

The range of upheaval has been recorded as high as 12 in (30 cm). [Conversely, settlement caused by soil moisture loss seldom exceeds 2 to 3 in (5 to 7.5 cm).] The general statement that upheaval is caused by water is true. However, water introduced into an expansive soil does not always cause upheaval. If the in situ soil moisture is near the PL, much of the soil's swell potential may have already been exhausted. Figure 7.2 gives the relationship between free swell and moisture content. The swell potential decreases dramatically as moisture content increases. (For these tests, a confining resistance equal to the assumed overburden was used.) At some point, additional moisture produces little or no swell.[15–17,86]

8.3.3 Redistribution of Existing Soil Moisture

Redistribution of existing soil moisture can occur to some extent, particularly when dealing with slab foundations. The presence of the foundation slab prevents, or inhibits, both the evapotranspiration and the penetration of water. This, in itself, conserves moisture and tends to encourage upward capillary flow. Given time, the bearing soil confined beneath the foundation is likely to increase in moisture content.[7,8] This increase will not be centralized but will be essentially constant over the expanse of the covered soil.[15,16,42,52,79,102,103] The vertical permeability in a clay soil is often approximated to be in the range of less than 1 ft per year.[65,99] The horizontal permeability is frequently assumed to be at least 10 times that, or 10 ft per year.[65,86,99] This means that lateral water flow can occur 10 times more readily than vertical flow. There is a very serious doubt that the natural accumulation of soil moisture beneath a slab foundation has ever been serious enough to require foundation repair. Certainly, in the author's experience, spanning nearly 35 years and well over 20,000 residential repairs, no such instance has ever been documented.

Various sources for natural subsurface moisture include wet weather aquifers, perched water zones, ponds, springs, and/or shallow water tables. As long as the bearing soil moisture content remains constant, little or no differential soil movement is anticipated. However, cyclic water availability regardless of source often produces soil movement—swell or shrink. Water intrusion into shallow soils can cause the upheaval described above. Natural water drawdown can result in settlement. The latter can occur as the result of draining or lowering the water level in surface ponds or subsurface aquifers. For the former, the soils are often nonexpansive. When subsurface water exists, the problem is best resolved prior to construction, and the practical solution is to initiate whatever steps are necessary to level the availability of water.

In some instances, chemical stabilization of the expansive clay might present a useful alternative (see Chap. 7).[16,17] Most chemicals are designed to abate swell and have, at best, limited capacity to influence moisture loss or settlement.

8.3.3.1 Summary of Center Doming.   The relationship of edge or center lift to describe potential foundation slab failures was introduced in the mid- to late 1970s.[90] From design concerns the approach seems technically sound. Problems can arise, however, through interpretation and application of the failure modes to actual foundation failure.

Edge lift rarely becomes a significant problem in foundation repair—although it remains a viable concern to design. Classically this phenomenon occurs when the soil moisture beneath a segment of the perimeter beam exceeds that beneath the interior slab. Heave of a perimeter beam is somewhat rare due in large part to the structure load transmitted to the beam. In residential and other lightly loaded construction, the loads imposed on the beam might be on the order of 600 lb/ft^2 (2900 kg/m^2) as opposed to 60 lb/ft^2 (290 kg/m^2) on interior floors. The added weight (load) substantially inhibits heave.[20] Further, the drastically reduced (or absence of) resistance (or load) on either side of a beam encourages the heave to be directed toward the path of least resistance. Areas adjacent to the beam might heave, but the loaded beam would be resistant to heave. Whatever the reason, problems associated with perimeter beam heave, due to water, are rare. (Tree roots have been known to heave a perimeter beam, particularly on pier and beam foundations.) Center lift actually refers to two different modes of failure: (1) perimeter settlement or (2) center lift, or as it is sometimes called, center doming.

Perimeter settlement is a concern and certainly should be an issue in foundation designs. However, foundation problems related to moisture losses in expansive soils are clearly overstated.[16,20] Soil moisture removal is slow, limited in scope, and easily abated if not reversed. When problems do develop, the remediation is both relatively inexpensive and easily handled. In fact, of all foundation repairs actually performed, less than 30 percent are the result of settlement and therefore are generally less costly.[16]

Center lift is a serious concern to both designs and repair. The fact of center heave is not in question. The principal cause for foundation repair is, in fact, upheaval. The cause for upheaval is generally beyond question. The presence of water beneath the foundation is generally accepted to be the cause. However, there are diverse opinions as to the source of the water that causes the soil to swell.

"Natural" Center Doming.   A few geotechs and engineers (mostly hired by insurance companies) advocate the natural occurrence of "center doming" beneath slab foundations on expansive soils. A few seem to believe that virtually all slab foundations will experience "center doming" given sufficient time. The theory is that soil suction will increase the moisture within the central area of the soil confined by the foundation to the point where soil swell and foundation heave is inevitable. First, for this to come about, the soil moisture in deeper soils must initially be higher than that for shallow soils. Soil suction and perhaps elevated temperatures could then make water migrate to dryer surface soils, resulting in soil swell. (The shallow soils are of primary concern because these soils have the greater propensity for volum-

metric movement.)[36,65,86] However, the soil permeability in the vertical direction is less than one-tenth that in the horizontal direction. Tom Petry has given an estimate of horizontal permeability (K_H) to be on the order of 10 to 20 ft per year.[6] Others have given a broad value for vertical permeabilities of 0.1 to 1.0 ft per year (10^{-7} to 10^{-6} cm/sec) for heavy clay.[20,65,86] (Increased K_H over K_V is due largely to roots, fissures, fractures, and/or normal sedimentary planes.) The disparity between K_V and K_H would seem to preclude the "dome" theory. A moisture buildup within the confined soil could readily occur, but the moisture content would be distributed reasonably uniformly over the entire surface soil—not as a "dome."[16,52,53,78,102] This theory would prevail even if interior perimeter soils were to lose moisture to the exterior. Second, bear in mind that (1) The natural doming wouldn't be dependent on or influenced by other moisture sources. For example, the mere process of repairing a leaking sewer system, would have no influence on arresting the heave. (2) All foundations of similar design, with other factors being equal, would eventually suffer the problem, and this simply does not occur. (3) Based on the extremely low permeabilities involved, natural doming, if possible, would require a long period of time to develop—perhaps in excess of 10 years. How then could this theory explain upheaval in foundations 2 to 10 years old? Reasonable design concerns can, however, resolve the problems of center heave as defined in the foregoing paragraphs.

Center Heave Due to Extraneous Water. One source of reliable data suggests that center doming is in fact a particularly serious problem to the stability of concrete slab foundations. The heave of concern involves the introduction of water beneath the foundation. This water can originate naturally (precipitation accumulated by bad drainage) or from domestic sources (plumbing leaks—supply or waste). Water produced naturally can be alleviated by conventional drainage improvements. The hidden "cancer" represented by domestic leaks, in particular the silent sewer, is the foremost problem. Perhaps the only reliable data concerning foundation repair identifies sewer leaks as the cause for at least 70 percent of all slab repairs. These data point out that once the cause of the problem (i.e., sewer leak) has been eliminated, movement arrests in about 90 percent of the documented causes. (In the remaining 10 percent, the continued movement was traced to an undiscovered plumbing leak in over half of them.)

A few engineers and geotechs completely discount plumbing leaks as a cause for foundation failure. Generally, these folks identify settlement (usually due to root or natural doming) as the cause for the failure. Again most of these proponents work exclusively (or nearly so) for insurance companies. Although these investigators deny plumbing leaks to be the cause of the problem, all agree to immediately repair the leak even though such repairs are typically quite expensive. To put the repair costs into some perspective, consider the example in Table 8.1. The foundation repair cost on this same job was proposed at $4908. The insurance company paid for the sewer repairs but denied responsibility for the foundation claim.

These same proponents also seem to state that a sewer leak does not produce sufficient water to cause a serious heave. Without going into the absurdity of this position (based on elementary soil mechanics or engineering, plumbing design, Manning equation, and/or common sense), it merely serves to note that when the sources for moisture are eliminated—movement ceases. Obviously the sewer leaks did contribute to (if not cause) the problem. All other arguments seem to suddenly become moot.

The cause for the foundation failure does not influence the pocketbook of the reputable repair contractor, who gives a repair estimate without concern for who pays the bill. The contractor does have responsibility to customers and should carefully inform them about the cause of the distress. This is simply because the contractor knows that if the original problem is not eliminated, the problem will recur. If the contractor correctly identifies the cause and this is not eliminated subsequent to repair, the contractors' warranty is likely null and void. The consumer is the party who suffers most. As a rule the insurance companies are the only parties affected by the cause of the problem.

8.3.4   Restoration

Upheaval represents that condition where some area of the foundation is distorted in a vertical direction while other areas remain as built. Remedial action involves raising the lowermost (unaffected) areas to a new, higher elevation in an effort to "feather" the crown, or high area. Seldom is the end product perfectly level, if such a concept exists. However, the consolation lies with the fact that the foundation was never level. The repair procedures are those described in Chaps. 2, 3, and 5, dependent upon whether the foundation is slab or pier and beam.

TABLE 8.1   Typical Costs for Sewer Repair (Slab Foundations)

Description	Quantity	Unit cost, $	Estimated cost, $
Dig and fill access for tunnel; includes pump protection	1	600	600
Tunnel under foundation	26	150	3900
Install pipe, fitting, and connections under slab materials plus labor	1	250	250
Backfill tunnel area	26	75	1950
Clean up		100	100
Permit/fees	1	100	100
Total			6900

8.3.5 Possible Recourse

Possible recourse due to foundation upheaval could include those cited in Sec. 8.2.4. However, conditions of upheaval might also be subject to insurance coverage. Again, as with Sec. 8.2.4, it is wise to consult an attorney experienced with this type of law. Until about 1996 many homeowner's B insurance policies contained coverage "for accidental discharge of water and damage resulting therefrom." This is, of course, not verbatim, but the intent is clear. However, claims under this coverage were often denied.

Any act by others that subjects your residential foundation to water (particularly a slab) might constitute "negligence." This could be a neighbor changing his or her drainage and adversely affecting your property. It might be flooding caused by a broken city water line. An improperly installed sewer or water supply system might also represent culpable acts. For the same reasons as those stated in Sec. 8.2.4, for preowned properties, sellers, realtors, engineers, and/or inspectors can also be held accountable for negligent acts.

8.4 FOUNDATION PROBLEMS NOT MOISTURE RELATED

8.4.1 Lateral Movements

Serious soil movement can be the result of some variety of lateral displacement. *Lateral movements* can occur because of soils eroding, sliding, or sloughing. This is generally associated with construction on slopes that have unstable bearing soils. Movement is often precipitated or exacerbated by the intrusion of water into the soil to the extent that both cohesion and structural strength are threatened or destroyed. California mud slides are one prime, though extreme, example. The structural damage caused by lateral movement is often complete destruction. When remediation is possible, the situation is addressed in a manner generally consistent with that for acute settlement. Severe lateral movement is generally beyond the methods available to the repair contractor. However, the contractor can often provide measures to stop lateral movement. This could involve the placement of retaining walls, earth anchors, terraces, or other such measures. Lateral movement is generally beyond the scope of this book.

8.4.2 Consolidation or Compaction

Settlement (not relative to soil moisture per se) can also occur as a result of consolidation or compaction of fill, base, or subbase materials. With respect to residential construction, the most common problem deals with construction on either abnormally thick or a sanitary fill. In either case, over time, the intended bearing soils fail due to consolidation. Normal settlement of fill is often active for periods up to 10 years—somewhat dependent upon the cycles of precipitation and drought. Sanitary fills can be active for longer periods of time due to voids continually provided by the decay of organic materials.

Consolidation of nonexpansive soils can result from the removal of pore water. One example of this might be instances where the water levels of lakes or ponds are lowered, allowing water to drain from the surrounding soils—often sand or coral. Elutriation of soluble material (usually salts) from soils by the invasion of water can create voids that at some point collapse and cause consolidation. In many cases, deep grouting is a required remedial procedure for the correction of either problem (see Chap. 7).[16] Once the deep-seated cause has been addressed, procedures common to foundation settlement can be used to "relevel" the structure.

8.4.3 Frost Heave

Frost heave occurs as a result of soil water freezing with sufficient expansion to cause the foundation member to heave or, in some cases as with basement walls, to collapse inward. Porous and permeable soils in subfreezing climates are the most susceptible to this problem. Frost heave can best be addressed in new construction design.

Frost heave in existing structures can result in slab heave or, in the case of basements, collapsed walls. Solutions to both problems should be very carefully planned, with the first step being to consult a geotechnical engineer who is skilled with this particular problem as well as the geographic location. Refer particularly to Sec. 8.3.4 and Chaps. 3 and 5.

8.4.3.1 Basements and Foundation Walls. Failure in basement walls occurs in areas with or without expansive soil and is often preponderant in colder climate regions. Where concern for frost lines exists, the propensity toward construction with basements is enhanced. As far as expansive soils are concerned, Colorado is one principal area where basements are common. As stated earlier, frost heave can exert sufficient lateral load on a basement wall to cause inward collapse, much the same as hydrostatic loads (expansive or nonexpansive soils) or the expansion of expansive soil. Regardless of the cause, the repair approach would be similar. Refer to Chap. 6 for details.

It might be interesting to note that above-grade foundation walls (with builtup floor systems, that is, dock high) suffer failures similar to those described in this section, except that the foundation failure (rotation) is toward the exterior and temperature is not a factor. Repair or restoration procedures are often much the same as those described in the following examples. As a rule, the retaining or aligning procedures are installed on the external side of the foundation wall where property lines permit. Refer to Chap. 6 for details on repair.

8.4.4 Permafrost

Permafrost is a reverse problem to frost heave. Both require the same conditions (i.e., prolonged very cold climate and lenses of water trapped within the bearing soil). The end result to the melting permafrost is subsidence. Unless the permafrost melts, there is no problem (foundation related). The design offered in Chap. 1 has been reported to be a workable, preventative measure. Once the problem has occurred, a possible solution might be deep (intermediate) grouting as discussed in Chap. 4. This approach could be successful, if the free water is eliminated. Again, the first and best option is to consult a local geotech engineer familiar with the specific problem, whether it is a design or remedial measure.

8.4.5 Construction Defects

Construction practices or mishaps often create conditions that are conducive to foundation problems. Where possible, these factors are grouped for either slab or pier and beam foundations.

8.4.5.1 Slab Foundations. There seems to be a number of events, inherent to the original construction, that can possibly result in foundation problems or in other cases impede the desired repairs. Some of these are

1. *Utility leaks* beneath the foundation (refer also, to Sec. 8.3).
2. Pouring the *slab foundation off grade.* Figure 8.6 presents grade elevations at various points over the area of the foundation. Note the differential of $-4^5/8$ in (-12 cm). Interior surfaces of the dwelling show only minor cracks in sheetrock and only a few doors are out of plumb. If leveling repairs were attempted based on these elevations, the structure would suffer severe distress. In fact, the extent of true *differential* movement noted in this foundation is "nominal" by most standards (see Sec. 10.1). At least part of the problem with this example originated because the brick ledge was used as a "benchmark." This is not acceptable. The brick ledge is most often off grade. In fact, brick masons use the first few courses of brick to attain a level mortar joint. Elevations should be taken from the mortar joint on top of the fifth brick course. Reliance on faulty information encourages errors in judgment and perhaps frivolous litigation.
3. *Faulty slab design or construction.* The slab foundation has been poured with (a) insufficient slab and/or beam thickness, (b) undersized, improper placement or absence of reinforcement, (c) *too much water,* which results in poor-quality concrete with substantial loss in strength. Not only do these defects encourage foundation problems, but they also hamper (if not prevent) proper repair.

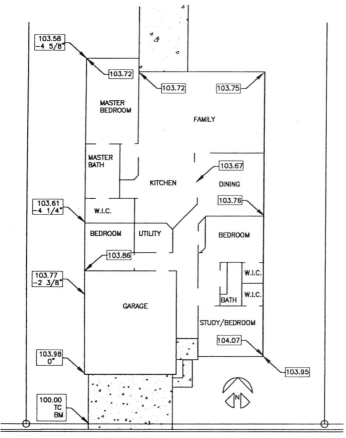

FIGURE 8.6 Foundation elevations.

4. *Add-on slabs* poured in contact with another slab that suffers from differential deflection. This situation is almost unrepairable. The common joint poses a real problem. If the faulty slab is raised, the add-on will be low. If the add-on slab is also raised, the existing framing will be destroyed. A rule of thumb: Never construct an add-on unless proper remediation has restored the original foundation. This situation affects slab and pier and beam foundations alike.

5. *Faulty exterior grade,* nonconsistent watering practices, and location and design of landscape plants each can promote foundation problems (see Chap. 9).

8.4.5.2   Pier and Beam Foundations   Foundation problems can also occur with pier and beam design. In fact, foundation problems happen on the same frequency as with slabs, but the degree of the problem is much less. The crawl space provides access for correcting minor grade problems experienced by interior floors. Other factors not so easily addressed include

1. *Limited or no crawl space.*   See Sec. 2.5.

2. *Insufficient ventilation.*   Refer to Sec. 2.5.

3. *Water collected in crawl space.*   This problem requires both proper drainage at the perimeter beam to prevent accumulation of excess water and adequate ventilation to control normal amounts of water. This is particularly important when the pier and beam foundation is of the "low profile" design (see Chap. 1).

4. *Warped wood,* particularly affecting the joists and/or girders, can be a serious deterrent to "leveling" attempts. When the wood substructure is subjected to a prolonged distortion (particularly in the presence of moisture), the individual wood members are subject to warp. If the warp is severe or well set, it is not likely that this condition can be remedied. That is, "leveling" by reshimming existing pier caps (or even adding supplemental ones), is not likely to produce "level" floors. Some improvement is generally possible but compromise is required.

5. *Faulty placement or design of piers and/or pier caps.*   In some cases, deficient materials were used to support the wood superstructure. This is particularly pronounced in older foundations where wood "stiff-legs" were used to support the foundation. These members are subject to deterioration as a result of decay and insect infestation. In modern practices, it is not particularly uncommon for a pier and pier cap to be located such that the girder is not properly supported. Sometimes the pier cap may be off center, tilted, short, or miss the girder altogether.

8.4.6   Negligent Maintenance

Negligent maintenance could refer to instances where

1. Standing water is permitted to approach and/or invade the foundation. Moisture accumulated in the crawl space (pier and beam) tends to not only acerbate wood warp or rot but also encourages instability of interior piers and pier caps. A substantial warp will prevent proper leveling (shimming pier caps) and cause increased costs. Water beneath slab foundations frequently causes upheaval.

2. Problems caused by erosion are not addressed before they become critical. This concern would also include embankment failure.

3. The neglect of proper maintenance procedures is probably the foremost issue in the category.

Proper maintenance is fully discussed in Chap. 9.

CHAPTER 9
ESTIMATING

9.1 INTRODUCTION

Estimating job costs forms the basis for any proposal. To establish these costs the following factors should be considered:

1. *Type of foundation.*   Pier and beam, slab, post-tension, slab on piers (are piers tied into perimeter beam?), etc. Are foundation plans and/or geotechnical data available?

2. *Cause and extent of the problem.*   Settlement? Upheaval? Upheaval is more costly to repair. Cause *must* be identified and eliminated to preserve repairs.

3. *Type, number, and placement of underpins.*   What access is available? (Access dictates what equipment can be used for drilling. Will concrete need to be broken out? Refer to Secs. 5.2 and 5.3.)

4. *Amount of mudjacking required.*   This is calculated on the basis of average raise times the area to be raised. Refer to Table 3.1.

5. Is water available at job site?

6. Is an unobstructed work and setup area sufficiently close to accommodate equipment? In most cases the pump/mixing equipment for mudjacking should be within 150 ft (45 m) of the farthest injection site. Without access, expensive alternatives must be considered. The latter could involve stage pumping (multiple pumps) or a very high water-to-solids ratio.

7. If mudjacking involves extensive inside pumping, such as the case might be with a warehouse slab, other questions arise such as (a) Is the slab doweled into the perimeter beam? If so, chances are the dowels must be cut to enable raising the slab at the perimeter beam. (b) Can mudjacking equipment be moved inside to facilitate access. Will exhaust fumes and dust become a hazard? (c) Are windows or doors available to route the pump hose to the work site? If the exterior walls are CMU (concrete masonry units), can access holes be created through the walls?

8. Is there sufficient work access in the crawl space? Is the area dry? A negative answer to either question can be costly. Refer to Chap. 2.

Most of these issues are determined based on an inspection report. A typical example of such a report is provided in Fig. 9.1.

The report in Fig. 9.1 represents the heart of estimating. It provides the readily available information and observations to permit an overview regarding the probable cause (or contributor) of the problem and at the same time provides a detail of the issues affecting the appropriate repair. Each and every mark on this paper is significant. For example, notice the comments regarding the east and west brick walls—"mtr.jts.reas.st." This translates to "mortar joints reasonably straight." This observation, taken into account with other factors such as the 2-in (5-cm) crown in the interior slab, helps identify upheaval as the culprit. The arrows, circles, and fractions at the corners indicate the relative movement at those spots; if no fraction is given, the movement is less than $1/4$ in (0.6 cm). The X indicate locations for piers. The X ? represents a location that might require a pier but will probably respond to mudjacking.

Prior to repairs, the plumbing test was conducted. A substantial leak was detected in the master-bath shower-commode area. The leak was repaired. From the inspection report it can be learned, among other things, that the foundation repair to the slab will require the installation of 25 drilled piers (12 in, or 0.3 m) plus 2 days of mudjacking. Access is available in 22 locations to permit use of the regular tractor (Bobcat or Case) -mounted drills. Three piers at the covered patio will need to be drilled by a limited-access rig. Water for the grout is available at the site, and the maximum length of grout hose is less than 100 ft (30 m). It is necessary to break out concrete at three locations for pier access. Based on this, a typical bid might be $10,411. This is an exceptional cost, necessitated in general by upheaval. Refer to Secs. 5.2 and 3.4.

The following sections and Chap. 3 provide the basis for specifying the 2 days of mudjacking. As an aside, there are two concerns when utility leaks are detected beneath a slab foundation. If the leak is detected *prior* to initiation of foundation repair, it is often considered prudent to postpone the repair procedure for several months to allow the bearing soil moisture content to reach some degree of equilibrium. This precaution might circumvent the

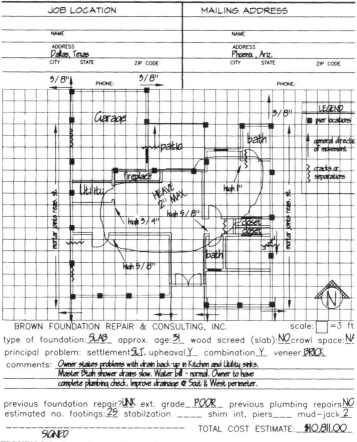

FIGURE 9.1 Inspection report example.

need for the contractor to remudjack the interior slab at some future date. The second issue involves detection of the leak *after* repairs are under way. At this point it is usually better to repair the leak and continue with the completion of work. This is particularly true if the perimeter has already been raised or is well along to being raised. The possible damage to an unsupported slab is more of a concern than the possible need to remudjack the slab at a later date. In either event the problem is of more concern to the contractor than the consumer. The contractors' warranty protects the consumer.

9.2 THE CASE FOR AND AGAINST USING GRADE ELEVATIONS TO ESTABLISH AMOUNT OF RAISE NEEDED

Note the designated movement noted on Fig. 9.1. This gives guidelines concerning how raising can restore the foundation to as-built condition without creating undue "new" damage. The actual measurement of this differential movement is not normally reflected in grade elevations. The latter reflects the contour of the foundation *at the time the measurements were made.* They do not necessarily reflect true movement. For grade elevations to be useful there must be at least two sets taken over a period of time. The sets of elevations should use the same benchmark (assumed zero) and should probably be timed at least 6 months apart. Elevations taken at the time construction was first completed are always useful to compare with later readings. For example, refer to Fig. 9.2. The grade elevations could suggest foundation movement in the range of $7^3/8$ in (19 cm). An inspection report found evidence of *differential* movement of less than 1 in (2.5 cm). If one attempted to raise this foundation 7 in (or even $4^1/2$ in), devastating destruction would result. The elevations and the inspection report each serve as a useful tool to identify upheaval. This is very important because (1) the cause of failure influences repair procedures, and (2) the cause of foundation problems must be identified and *corrected* concurrent with any repairs; otherwise, the repairs cannot expect to be permanent. Because of the preponderance of upheaval in slab repairs, this importance is even more emphasized. This is discussed further in Sec. 9.3. For one reason or another some individuals, for self-serving reasons, have warped notions regarding the occurrence of upheaval, sometimes claiming that thousands of gallons of water are required to cause significant foundation heave. Others claim that "once the source of water is eliminated, the foundation will correct itself." Neither of these positions makes sense. Consider Sec. 9.3.

9.3 SOIL SWELL (UPHEAVAL) VERSUS MOISTURE CHANGES

Some discussion has already been devoted to the issue of soil swell. Tucker and Davis presented data for a particular soil wherein a 4 percent increase in soil moisture was sufficient to cause a $1^1/4$-in (3.2-cm) vertical rise in a type B, FHA slab foundation, with a resulting potential swell force of 9000 lb/ft^2 (43,900 kg/m^2). This same study produced data that suggested a soil "active" depth of about 7 ft (2.1 m) exterior to the foundation. However, these same data showed that over 85 percent of the total soil moisture change occurred in about the top 3 ft (0.9 m).[15,17]

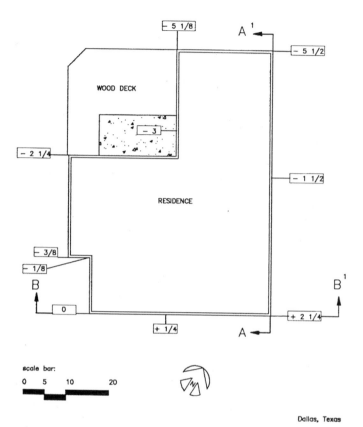

FIGURE 9.2 Relative elevation survey.

What can all these seemingly unimportant factors tell about soil swell? First, refer to Fig. 9.3a. Here it is assumed that a cubic foot of soil is isolated from the surrounding soil. Assume an imaginary glass box 1 ft^2 by 3 ft (0.3 m^2 by 0.9 m) deep filled with the soil described by Dr. Tucker, which existed at an initial moisture content of 20 percent with a final moisture content of 24 percent. The 9000 lb/ft^2 (43,900 kg/m^2) swell potential of the "confined" soil will tend to raise the lightly loaded slab [approximately 60 lb/ft^2 (290 kg/m^2)] over an area large enough to counter the upward thrust. In this case 9000 lb/60 lb/ft^2 = 150 ft^2 (13.5 m^2), or roughly an area of 12 ft^2 (3.6 m^2). This analysis admittedly takes certain liberties but is nonetheless technically valid. The *resistance* to heave would be materially influenced and enhanced by such factors as steel reinforcement, cross beams, or load-bearing walls. However, the intent here is not to confuse but to illuminate. The resisting moment of a beam is a square function of its depth (steel reinforcement not considered); for example, under identical conditions a wood beam 2 × 4 in (5 × 10 cm) on edge will support about 4 times the load it would if laid flat. When steel rebar within the concrete beam is considered, the moment of inertia (or stiffness) varies as the cube of the depth. In this case, doubling the depth of the beam would increase the rigidity by a factor of perhaps 8.[37,95]

To pursue this train of thought, how much water is actually required to produce the 4 percent increase? Assume $W_s + W_w = 100$ lb/ft³ (1602 kg/m³); $W_s = 86$ lb/ft³ (1376 kg/m³); initial moisture, W% = 16, $W_w = 14$ lb/ft³ (224 kg/m³); final moisture, W% = 20, $W_w = 17$ lb/ft³ (272 kg/m³); where W_w is weight of water, W_s is weight of soil, and W% = W_w/W_s. All values are based on a 1-ft³ (16-kg/m³) sample. Based on this, the added weight of water would be $W_w = 17 − 14 = 3$ lb/ft³ (48 kg/m³), or 0.36 gal (1.4 L) per cubic foot. Assume the constraints set forth by Fig. 7.13. In the case at hand, this would approximate only 1.2 gal (0.4 gal/ft³ × 3 ft³) or 4.2 l/ft³ (150 l/m³). Again for simplicity assume that the source for water had preexisted for 12 months. Then the daily input of water (and the amount required to produce the 4 percent increase) would be only 143 drops per day.[15–17] For example,

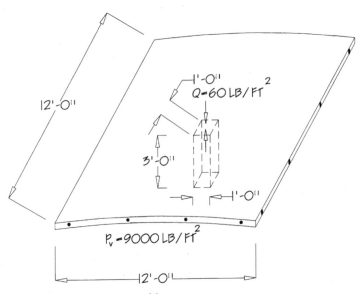

FIGURE 9.3a Heave of concrete slab.

1.2 gal/12 months $= 0.10$ gal/month $= 0.0033$ gal per day

$= 0.43$ oz per day

$= 14.3$ mL per day

$= 143$ drops per day

$= 0.10$ drop per minute

Admittedly, this example isolates the cube of wetted soil, and in real life that could not occur. However, the relative quantities show a clear picture. Also, as a broader area is wetted, the potential soil heaved proportionately expands. In other words, if the wetted area expands laterally, the heaved area of the slab expands almost in direct proportion, although the magnitude of the heave could be lessened.

As the soil expands, what happens to the slab? First, assuming a conventional monolithic deformed bar slab, the steel is stressed to a distorted length, which is not going to recover without some form of reverse stress. Along these lines, also consider Fig. 9.3b. If this slab is heaved by 4 in (10 cm), each steel rebar within the 12-ft (3.6-m) heaved area will be stretched by $1/4$ in (0.6 cm). Note as well that the elongation of the rebar will be anything but uniform. (With post-tension slabs, where the cables are sleeved, this would not be the absolute case.) Even after the cause of the heave is alleviated, the domed area is not likely to return to a level (or near level) condition. In fact, experience dictates that the distressed area will not improve materially unless appropriate remedial actions are initiated.

Where can the water come from to cause the swell? This subject has been discussed to some extent in earlier paragraphs. Sewer leaks represent the most prevalent source. Figure 9.4 depicts a sewer line with a separation. [The normal, minimal gravity fall in the sewer pipe is $1/8$ in (0.3 cm) per linear foot—approximately 1 ft/100 ft (0.3 m/30 m).] Wastewater directed into the sewer forms a vortex (turbulent flow), which creates centrifugal force and tends to throw liquid from the pipe if any separation exists. Eventually, the flow settles down to the laminar regime, with the major velocity being down the pipe centerline. In laminar flow, the amount of water leaking from the pipe might be lessened. However, as shown in preceding paragraphs, very little water is required to cause a potentially serious threat.

Not all expansive soils swell when subjected to available water (see Fig. 7.3). If the existing moisture for these particular soils is above 24 percent, virtually all capacity for swell has been lost. Also, slab heave will not always appear to be fairly uniform as depicted by Fig. 9.3a. The figure only illustrates a principle of force versus resistance and borders on an ideal condition. (For example, the wetted area is assumed to be at the surface. Often the affected area is at the bottom of the plumbing ditch and could be several feet below surface. Also the presence of porous fill and/or subsurface contact with foundation beams

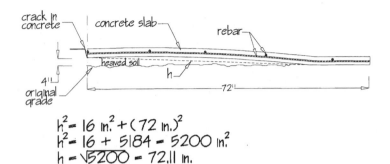

$$h^2 = 16 \text{ in.}^2 + (72 \text{ in.})^2$$
$$h^2 = 16 + 5184 = 5200 \text{ in.}^2$$
$$h = \sqrt{5200} = 72.11 \text{ in.}$$

$$\text{stretch} = (72.11 - 72.0)(2) = 0.22 \text{ in.}$$

or approximately 1/4" over a 12 ft. span
(this would involve each rebar in heaved area)

In conventional slabs, the stress on rebar will not likely be uniform but generally localized to ares with concrete cracks.

In pt slabs, the stress will be more uniform to the cable with slab cracks more random.

FIGURE 9.3*b* Rebar stretch versus heave. Assume that the heave depicted is 4 in (10 cm); the stretch in the rebar would approximate 0.22 in (5.6 mm). Again, this is an idealistic representation, but it serves to illustrate an example.

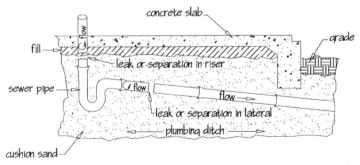

FIGURE 9.4 Residential sewer.

can influence the pattern of water flow. The latter would cause the heave to take an elongated rather than circular pattern. In many cases, particularly with older properties, proper mudjacking eliminates these flow channels.)

Realistically, the deformity of a concrete slab foundation would appear more as Fig. 9.5 suggests than as shown in Fig. 9.4. The added structural load on the beam coupled with the increased resistance to deflection provided by the beam strength distorts the doming effect. In effect, the slab resembles a quilted surface, except that the individual cells need not be of organized dimension. A topographical view of such a slab might resemble Figs. 9.6 or 9.7.

Figure 9.6 includes elevations and has indications of minor settlement at the northwest, west of entry, and possibly the southeast corners. The major movement is center slab heave in the shaded area. Figures 9.6 and 9.7 also serve as foundation inspection field drawings intended to provide information sufficient for a repair estimate. The heaved area is generalized from observation of differential movement.

Section 9.2 provides additional discussion regarding the difference between grade elevations and differential movement. Little, if anything, can be done to improve variations in grade elevation caused by initial construction. In fact, attempts to do so are likely to cause serious additional distress. Refer again to Fig. 9.2. It would be impossible to raise areas of this foundation to the extent the elevations suggest.

The foregoing will help provide the background information necessary to prepare a workable estimate. The appropriate repair can be balanced against the cause. Do all cases of foundation movement warrant repair? Who decides at what point repairs are feasible? Section 9.4 addresses these issues.

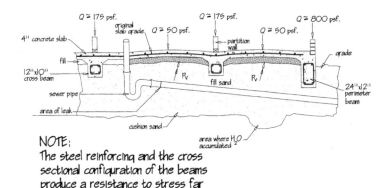

FIGURE 9.5 Example of slab displacement due to upheaval resulting from soil swell.

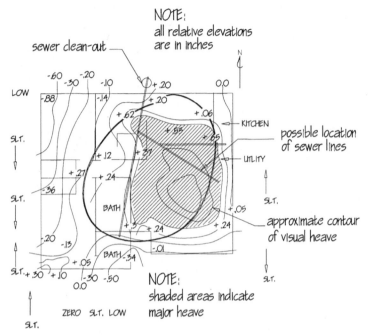

FIGURE 9.6 Differential elevations in a slab foundation.

9.4 HOW IS THE NEED FOR FOUNDATION REPAIR ESTABLISHED?

What extent of movement is sufficiently serious to warrant or demand repairs? There is no uniformly established rule. Several factors further complicate this issue: (1) Few, if any, residential foundations are constructed level, (2) a foundation being out of level does not normally render the property uninhabitable or unsafe, and (3) at least in part because of the foregoing, virtually all approaches to foundation repair (leveling) include some degree of compromise. The net goal is a foundation of "tolerable" appearance and behavior. The problem lies in the conceptual definition of *tolerable*.

In attempting to establish a reasonable basis for defining the need and scope of foundation repair, the preferred approach is to provide the most acceptable appearance along with the most stable foundation, at the least possible cost. Determining the point of movement at which foundation repair is demanded is equally elusive.[65,73] Because most foundation repairs are performed as an aesthetic choice rather than a true structural necessity, the mental attitude of the property owner becomes a prime factor. Other factors that play a part are property age and value, spendable income of the owner, peer pressure, cost of foundation repair versus continued cosmetic repair, the movement is ongoing, cause of distress (settlement versus upheaval), and the likelihood that proper maintenance could arrest the movement. Before any foundation repair is performed, the cause of the differential movement must be identified and subsequently corrected. Otherwise, recurrent problems are likely, regardless of the foundation repairs. A semitechnical approach for establishing "tolerable" and "intolerable" foundation movement is developed in the following paragraphs.

Tensile strain for a typical mortared masonry wall is 0.0005 in/in (strain is determined by measuring deflection divided by distance). Thus, such a wall 8 ft (2.4 m) in height could resist movements up to about 1 in (2.5 cm) over 65 ft (20 m). On the other hand, a 1-in movement over 15 ft (4.6 m) could produce a single crack separation of about $1/4$ in (0.6 cm). Note the emphasis on "single crack" separation (see Fig. 9.8). Multiple cracks reduce separation width proportionately.

Lambe and Whitman present the relationship between strain and distance a bit differently.[65] Refer to Fig. 9.9. If L is replaced by x in b and by $x/2$ in c,

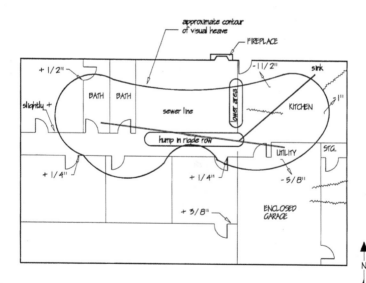

FIGURE 9.7 Heave contour determined by field drawing.

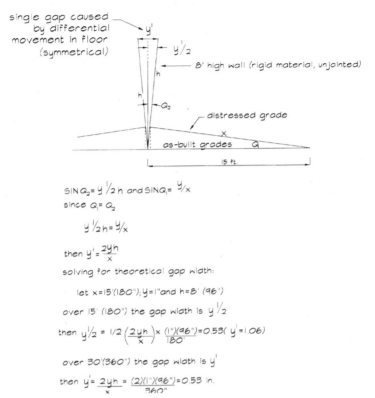

single gap caused by differential movement in floor (symmetrical)

y'

$y/2$

8' high wall (rigid material, unjointed)

h

h

Q_2

distressed grade

as-built grades

x

Q_1

15 ft.

$SIN\ Q_2 = y'/2\,h$ and $SIN Q_1 = y/x$

since $Q_1 = Q_2$

$$y'/2\,h = y/x$$

then $y' = \dfrac{2yh}{x}$

solving for theoretical gap width:

let $x = 15'(180")$; $y = 1"$ and $h = 8'\ (96")$

over 15' (180") the gap width is $y'/2$

then $y'/2 = 1/2 \left(\dfrac{2yh}{x}\right) \times \dfrac{(1")(96")}{180'} = 0.53\ (y' = 1.06)$

over 30'(360") the gap width is y'

then $y' = \dfrac{2yh}{x} = \dfrac{(2)(1")(96")}{360"} = 0.53$ in.

FIGURE 9.8   Theoretical single gap in 8-ft (2.4-m) wall.

the analysis is similar to that shown in Fig. 9.8. Consistent with this analogy, interior features such as door frames can tolerate movement on the order of 1 in (2.5 cm) over 12.5 ft (3.8 m). This is borderline. Increased movement usually causes the doors to be nonfunctional.

This discussion brings into focus the problems inherent in defining tolerable deflection. Various conditions suggest different values. The distance over which the deflection occurs is the paramount concern, followed by the nature of the structure. An arbitrary value for acceptable movement that is recognized by some repair contractors is crack separation in excess of $1/4$ to $3/8$ in (0.6 to 1 cm). These roughly equate to a differential movement of 1 in (2.5 cm) over 30 ft (10 m) or 25 ft (8 m), respectively. Normal construction often accepts a tolerance of as much as 1 in in 20 ft (6 m).[6] This again emphasizes the difference between possible as-built grade variation and differential movement.

Kirby Meyer suggests the values expressed in Table 9.1 as the criteria for foundation failure threshold.[73] In other words, movements that exceed the numbers indicated in the table suggest a failure condition that should be considered as both serious and warranting remedial attention.

Dov Kaminetzky approaches the situation from a slightly different perspective.[61] Table 9.2 presents his classification of distress based on visible damage. He defines moderate damage as occurring at approximate crack widths of $1/8$ to $1/2$ in (3.2 to 12.7 mm). This is the point at which remedial action is suggested. Overall, the acceptable magnitude of differential movement is 0.3 percent (refer also to Chap. 11).

A slightly different approach to a qualitative analysis for a slab-on-grade foundation is being considered by the Post Tension Institution. This directive is titled "Tentative Performance Standard Guidelines for Residential and Light Commercial Construction" as developed by the PTI of Phoenix, Arizona. This bulletin addresses new construction (type A) as well as existing structures (type B). Concerns herein are limited to the latter. The following presents the "point" basis for foundation evaluation for type B:

1. Foundation cracking

2. Sheetrock wall distress

3. Doors

4. Exterior cladding of brick, stone, or stucco

5. Separation of materials

6. Foundation levelness

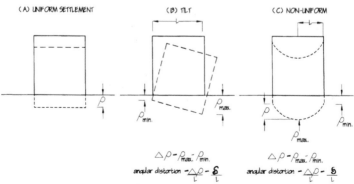

FIGURE 9.9 Type of settlement.

TABLE 9.1 Recommended Foundation Failure Criteria for Selected Building Maximum Allowable Values

Condition	Differential across two points, y/x*	Overall slope, % y/x*	Total vertical movement, in
Single or multifamily wood frame dwellings up to three stories			
Exterior masonry 8 ft high	1 in/40 ft		
Exterior masonry >8 ft high	1 in/50 ft		
Exterior plaster 8 ft high	1 in/40 ft		
Exterior nonmasonry	1 in/16.6 ft		
Interior sheetrock walls 8 ft high	1 in/30 ft		
Interior sheetrock walls >8 ft high	1 in/40 ft		
Interior wood-paneled walls	1 in/76.7 ft		
Small-span structures with cross-roof trusses	1 in/30 ft		
All		0.3 (1 in/30 ft)	4
Steel-frame building with metallic skin, no sensitive equipment, average range			
If isolated location with soft adjacent improvements and maximum drainage	1 in/16.7 ft	1.0	6
Concrete framed industrial building with CMU or masonry	1 in/30 ft		
High-exposure low-rise retail or office building with glass or architectural masonry walls	1 in/50 ft	0.2	2 (but check entry transitions)

*x is lateral distance over which vertical deflection y occurs. In the case of overall slope, x would generally be the length or width of the foundation, as the case might be, and y the overall deflection.
Source: After Kirby T. Meyer, "Defining Foundation Failure."[73] See also Refs. 60, 61, and 1.

TABLE 9.2 Classification of Visible Damage*

Extent of damage	Description of typical damage	Crack width, in (mm)
	Hairline cracks of less than about 0.005 in (0.13 mm) are classified as negligible.	
Very light	Isolated light fracture in building. Cracks in exterior walls visible on close inspection.	$1/64$ to $1/32$ (0.4 to 0.8)
Light	Light inside fracture visible on floors/partitions and on the exterior of buildings. Doors and windows may stick slightly.	$1/32$ to $1/8$ (0.80 to 3.2)
Moderate	Moderate cracks are visible on the inside and on the exterior of buildings. Doors and windows stick. Utility pipes and glass may fracture. Water and air penetration through exterior.	$1/8$ to $1/2$ (3.2 to 12.7)
Extensive	Severe cracks are visible on the inside and on the exterior of buildings. Windows and door frames are skewed and "locked in" noticeably. Walls lean or bulge noticeably; some loss of bearing in beams. Cracks in utility pipes and glass brick; water and air penetration through exterior.	$1/2$ to 1 (12.7 to 25.4)
Very extensive	Very extensive cracks are visible on the inside and on the exterior of buildings. Broken pipes and glass. Full loss of beam bearing; walls lean or bulge dangerously. Structure requires shoring. Danger of instability.	1 (25.4) or wider but depends on number and location of cracks

*In evaluating the degree of damage, consideration must be given to its location in the building and structure (e.g., points of maximum stress). Crack width is only one aspect of damage and should not be used alone as a direct measure of damage and refers to existing "old" cracks. New cracks must be monitored for possible increase in width. A criterion related to visible cracking is useful because tensile cracking is so often associated with settlement or movement damage. Therefore, it may be assumed that the state of visible cracking in a given material is associated with its limit of tensile strain.

Source: Dov Kaminetzky, "Rehabilitation and Renovation of Concrete Buildings."[61]

EVALUATION

ITEM	POINTS

1) Foundation cracking (maximum points allowed: 7)

    a) Number of distinct cracks

1–4	1
5 or more	2

    b) Size of largest crack

0 to $1/16$ in	1
$3/32$ to $1/4$ in	3
$5/16$ in or greater	5

Remarks: Corner wedge cracks are not considered distress cracks and score no points. Tight shrinkage cracks or hairline cracks too small for a business card to fit into also are not counted.

2) Sheetrock wall and cabinet distress (maximum points: 8)

    a) Number of distinct wall cracks or wall/cabinet separations within the interior of the house

1–5	1
6–10	2
11 or more	5

    b) Size of largest sheetrock crack or separation

0 to $1/8$ in	1
$3/16$ to $5/16$ in	3
$3/8$ in or greater	5

Remarks: Minimum crack length must be 6 in.

3) Doors (maximum points allowed: 8)

    a) Number of doors sticking or not latching properly

1–3	1
4 or more	3

    b) Any door totally inoperable — 5

Remarks: Door distress must be a function of foundation movement, not due to material moisture changes.

4) Exterior cladding of brick or stone veneer or stucco (maximum points: 8)

    a) Number of distinct brick veneer cracks

1–4	1
5–10	2
11 or more	3

    b) Size of largest crack

0 to $1/8$ in	1
$1/4$ to $1/2$ in	3
$5/8$ in or more	5

5) Separation of materials (maximum points allowed: 7)
   a) Number of locations where brick veneer and adjacent
      material are separated

1–4	1
5 or more	2

   b) Size of largest separation

0 to $1/8$ in	1
$3/16$ to $5/16$ in	3
$3/8$ in or greater	5

Remark: Normal caulk and wood shrinkage do not count.

6) Foundation levelness (maximum points allowed: 5)
   a) Slope at any 8-ft line

0 to $1/2$ in	0
$7/16$ to $3/4$ in	1
$13/16$ to 1 in	2
$1^{1}/16$ to $1^{1}/4$ in	3
more than $1^{1}/4$ in	5

Remarks: Elevation inconsistences representing a floor slope of greater than $1^{1}/4$ in in 8 ft is unacceptable.

Conclusion. Any structure with a total score of 25 points or more is determined defective, or "failed." Example: Foundation and Superstructure Quantitative Criteria. A single-story wood-frame brick-veneer structure with a post-censioned slab-on-grade foundation. The property was built in approximately 1990. The evaluation follows. The system will be used to grade the condition of a structure. The structure will be considered a type B because there were no prior slab elevation readings and the structure was existing. The system is quantitative and considers the structure to be defective, or to have failed, when a total score of 25 points is attained.

EVALUATION

ITEM	POINTS
1) Foundation cracking	
a) Number of distinct cracks	
3 distinct cracks	1
b) Size of cracks	
$3/32$ in	3
2) Sheetrock wall and cabinet distress	
a) Number of distinct cracks	
17 distinct cracks	3
b) Size of largest crack over 10 in	
$3/8$ in	5

3) Doors and windows
  a) Number of doors sticking
      3 doors and/or windows sticking   1
  b) Number of doors and/or windows totally inoperable
      3 inoperable doors/windows   5

4) Exterior brick cracking
  a) Number of distinct brick cracks
      6 distinct cracks   2
  b) Size of largest crack
      $^1/_{32}$ in   1

5) Separation of materials
  a) Number of locations
      4 locations   1
  b) Size of largest separation
      1 in   5

6) Foundation levelness
  a) Slope at any 8-ft line
      $2^1/_4$ in   __5__

Total Evaluation Points   **32**

The Worst Possible Score   48

From this analysis this foundation has failed. This method is sometimes more tolerable than any of the foregoing, even though more observations are included in the evaluation. To equate the PTI evaluation to actual repair needs, this property was later inspected by a foundation contractor to provide a repair estimate. [The engineer's 1997 evaluation indicated a maximum grade differential of "plus or minus $2^1/_4$ in" (6 cm), or about $4^1/_2$ in (12 cm).] The inspection drawing prepared by the contractor is given in Fig. 9.10. From this inspection, the magnitude of *differential foundation movement* is less than 2 in (10 cm). Note also that the principal problem is upheaval running essentially north to south between the two baths that had prior leaks. As noted on other elevation exhibits, the extent of grade differentials is often not a reflection of differential foundation movement but a statement of original construction. Note that the maximum differential is stated to be about $1^1/_2$ in (4 cm). In this case foundation repairs are recommended subsequent to repair of the existing sewer leak.

Home Owners' Warranty (HOW) uses a still different approach. Their booklet describes "major structural" defects as those which meet two criteria: "(a) it must represent actual damage to the load-bearing portion of the home that affects its load-bearing function and (b) the damage must vitally affect or is imminently likely to produce a vital effect on the use of the home

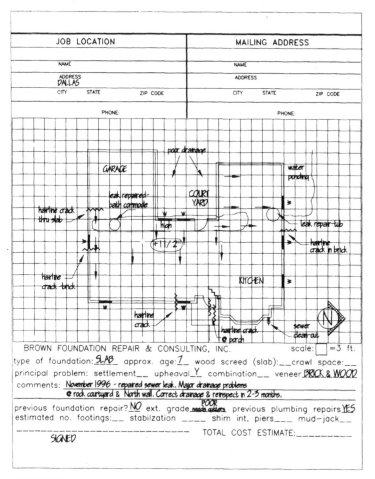

FIGURE 9.10 Inspection report.

for residential purposes." In HOW Chap. 3, "Concrete," the following are cited as excessive cracks: foundation or basement walls—greater than $1/8$ in (0.3 cm); basement floors—in excess of $3/16$ in (0.45 cm); garage slabs—wider than $1/4$ in (0.6 cm); foundation slab floor—wider than $1/8$ in (0.3 cm). Chapter 4, "Masonry," defines cracks in veneer as excessive when they exceed $3/8$ in (0.9 cm) in width. Chapter 6, "Wood and Plastics," defines as intolerable floors that are out of level more than $1/4$ in (0.6 cm) over a linear distance of 32 in (0.8 m). General floor slope within any room should not exceed 1 in (2.5 cm) over 20 ft (6 m). On the other hand, HUD (at least some regional offices) often considers a property uninsurable if the floor is off level more than $11/16$ in (1.7 cm) over 20 ft (6 m). This translates to 1 in (2.5 cm) over 30 ft (10 m).[18] Both HOW and HUD offer the homeowner some measure of insurance. HUD insures homeowner mortgage loans and HOW provides them with insurance protection against major structural defects. Warranty Underwriters Insurance Co. and Home Buyers Warranty are examples of other organizations that offer home buyers some degree of protection in the form of insurance against major household defects such as foundation failure. Each entity offers a somewhat different policy regarding insured losses.

Basic homeowners' insurance policies generally cover foundation repair only in cases where the foundation movements are caused by accidental discharge of water from the household plumbing system (sewer or supply).

If you have insurance and suspect a foundation problem, the best bet is to first contact a qualified engineer or foundation repair expert and follow the advice given.

CHAPTER 10
PREVENTIVE MAINTENANCE

10.1 INTRODUCTION

Preventing a problem always makes more sense than curing one. Chapter 8 discussed predominate causes for foundation failures. Chapters 2, 3, and 5 covered methods for correcting common foundation failures. Obviously, if the cause for the problem does not exist, neither will the problem. This chapter will focus on measures that encourage foundation stability, with a specific focus on those problems relative to expansive soils. Certain maintenance procedures can help prevent or arrest foundation problems if initiated at the proper time and carried out diligently. The following are specific suggestions on how this can be accomplished.

10.2 WATERING

In dry periods, summer or winter, water the soil adjacent to the foundation to help maintain constant moisture. *Proper* watering is the key and will be discussed in the following paragraph. Also, be sure drainage is away from the foundation prior to watering. Remember, too much water can cause far more problems than too little.

When a separation appears between the soil and foundation perimeter, the soil moisture is low and watering is in order. Water should not be allowed to stand in pools against the foundation. Never attempt to water the foundation with a root feeder or by placing a running garden hose adjacent to the beam. Both represent uncontrolled watering. Sprinkler systems often create a sense of "false security" because the shrub heads, normally in close proximity to the perimeter beam, are generally set to spray away from the structure. The design can be altered to put water at the perimeter and thereby serve the purpose. This is done by replacing the sector heads with strip heads. However, in the end, the use of a soaker hose is still often the best solution. In either event, watering should be uniform and cover long areas at each setting, ideally 50 to 100 linear feet (15 to 30 m). From previous studies of water infiltration and run-off, it is evident that watering should be close to the foundation, within 6 to 18 in (15 to 45 cm), and timed to prevent excessive watering.

Proper grading around the foundation will also prevent unwanted accumulation of standing water. Sophisticated watering systems that utilize a subsurface weep hose with electrically activated control valves and automatic moisture monitoring and control devices are reportedly available. The multiple control devices are allegedly designed to afford adequate soil moisture control, automatically and evenly around the foundation perimeter. Reportedly, the control can be set to limit moisture variations to ±1 percent. Within this tolerance, little if any, differential foundation movement would be expected in even the most volatile or expansive clay soils. The key to this system is a true and proven ability to control water output and placement.[15–17]

Avoid watering systems that make outlandish claims. They can often cause more problems than they cure. One example is the so-called *water, or hydro, pier*. One claim is that a weep hose with sections placed vertically into the soil on convenient centers (often 6 to 8 ft, or 1.8 to 2.4 m) will develop a "pier" that will then support and stabilize the foundation. The pier is allegedly the product of expanded clay soil (see Fig. 5.15). Some contractors even claim the system will raise and "level" foundations. Outside the other obvious deficiencies, this procedure seems to lack controls capable of either equating water added to in situ moisture or monitoring total soil moisture. Lack of performance here introduces several problems when dealing with expansive soils.[16] First, moisture distribution within the soil is seldom, if ever, uniform. Second, excessive soil moisture approaching the liquid limit (LL) can cause a soil to actually *lose* strength (cohesion). Third, although *consistent* moisture content in the expansive soil will normally prevent differential deflection of the foundation, the method under discussion is not likely to meet the requirements. And, fourth, water replenishment into an expansive soil will seldom, if ever, singularly accomplish any acceptable degree of leveling. (Minor settlement, limited in scope, could be the exception.[16,26])

Where large plants or trees are located near the foundation, it could be advisable to conservatively water these, at least in areas with climatic (C_w) factors below about 25.[36] (These areas are generally classed as "semiarid."[36] Lower C_w values lean toward being more arid.) As far as foundation stability is concerned, the trees or plants most likely to require additional water are those that (1) are immature, (2) develop root systems that tend to remove water from shallow soils, and (3) are situated within a few feet of the foundation. Refer also to Chap. 3.

10.3 DRAINAGE

It is important that the ground surface water drain away from the foundation. Proper moisture availability is the key. Excessive water is frequently detrimental. Proper drainage will help avoid excess water. Where grade improvement is required, the fill should be low-clay or sandy-loam soil. The slope of the fill need not be exaggerated but merely sufficient to cause water to flow outward from the structure. A 1 percent slope is equivalent to a drop of 1 in (2.5 cm) over approximately 8 ft (2.4 m). A satisfactory slope is often assumed to be 1 to 3 percent. Too great a slope encourages erosion. The surface of fill must be below the air vent for pier and beam foundations and below the brick ledge (weep holes) for slabs. Surface water, whether from rainfall or watering, should never be allowed to collect and stand in areas adjacent to the foundation wall.

Along with proper drainage, guttering and proper discharge of downspouts is quite important. Flower-bed curbing and planter boxes should drain freely and preclude trapped water at the perimeter. In essence, any procedure that controls and removes excess surface water is beneficial to foundation stability. Water accumulation in the crawl space of a pier and beam foundation is also to be avoided. Low-profile pier and beam foundations can be particularly susceptible to this problem. Drainage control and adequate ventilation serve as the best preventive measures. As a rule of thumb, vents should be provided on the ratio of 1 ft^2/150 ft^2 of floor space. Where construction design prevents an adequate number of vents, the desired ventilation can be implemented by the use of forced-air blowers.

Domestic plumbing leaks (supply and sewer) can be another source of unwanted water.[24,26,51,79,83,86] Extra care should be taken to prevent and/or correct this problem. Water accumulation beneath a slab foundation accounts for a reported 70 percent of all slab repairs.[15–17] Sewer leaks are responsible for very high percentage of these failures (see Chap. 2).

10.3.1 French Drains and Subsurface Water

French drains are required, upon occasion, when subsurface water migrates beneath the foundation. Figure 10.1 shows a typical French drain. When the foundation is supported by a volatile (high-clay) soil, intrusion of unwanted water must be stopped. The installation of a French drain to intercept and divert the water is a useful approach.[15–17,26,79] The drain consists of a suitable ditch cut to some depth below the level of the intruding water. The lowermost part of the ditch is filled with gravel surrounding a perforated pipe. The top of the gravel is continued to at least above the water access level and often to or near the surface.

Provisions are incorporated to remove the water from the drain either by a gravity pipe drain or a suitable pump system. Simply stated, the French

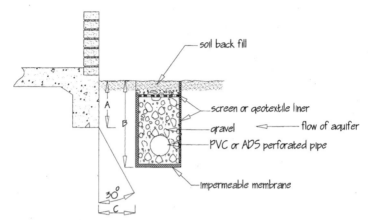

soil back fill

screen or geotextile liner

gravel ⟵———— flow of aquifer

PVC or ADS perforated pipe

impermeable membrane

A

B

30°

C

FIGURE 10.1 Typical French drain. Generally C is equal to or greater than A, and B is greater than A+2 ft (0.6 m). The drain should be located outside the load surcharge area.

drain creates a more permeable route for flow and carries the water to a safe disposal point. If the slope of the terrain is not sufficient to afford gravity drainage, the use of a catch basin/sump pump system is required. The subsurface water commonly handled by the French drain is perched ground water or lateral flow from "wet weather" springs or shallow aquifers.

Where the conditions warrant, the design of the drain can be modified to also drain excessive surface water. This is readily accomplished by adding surface drains (risers) connected to the French drain system. An alternative approach is to carry the gravel to the surface. (A proper drain intercepts and disposes of the water before it invades the foundation.) Water from downspouts should not be tied directly into the French drain; a separate pipe drain system is preferred. The second, solid, pipe could, however, be placed in the French drain trench.

A French drain is of little or no use in relieving water problems resulting from a spring within the confines of the foundation because it is almost impossible to locate and tap such a spring. This condition is also rare because a wet site causes real problems for the builder.

When distress problems exist prior to the installation of a French drain, foundation repairs are often required. In that event, these repairs should be delayed to give ample time for the French drain to develop a condition of moisture equilibrium under the foundation. Otherwise, recurrent distress (repairs) can be anticipated due to the disturbance of soil moisture introduced by the drain.

10.3.2 Water or Capillary Barriers

In an effort to maintain constant soil moisture, measures that impede the unwanted transfer of soil moisture can be considered. One such attempt is the use of moisture barriers. The barriers may be either horizontal or vertical, permeable or impermeable, as discussed below.

10.3.2.1 Horizontal Barriers. Horizontal, *impermeable* barriers can be as simple as asphalt or concrete paving or polyethylene film (see Fig. 10.2). These materials are used to cover the soil surface adjacent to the foundation and inhibit evaporation. Coincidentally, the covers could also restrict soil moisture loss to transpiration because vegetation would neither grow nor be cultivated in the sheltered area.

Permeable horizontal barriers usually consist of little more than landscaping gravel or granular fill placed on the soil surface previously graded for drainage. Moisture within the porous material cannot develop a surface tension, and no adhesive forces will exist, both of which are required to create a capillary (or pore) pressure. Gravity (drainage) then becomes the factor dictating free water movement.

10.3.2.2 Vertical Barriers. The vertical *impermeable* capillary (or water) barrier (VICB) is intended to block the transfer of water laterally within the affected soil matrix. Figure 10.3 shows typical vertical barriers. Placed adjacent to a foundation, the VICB should maintain the soil moisture at a constant level within the foundation soil encapsulated by the barrier.[1-4] As an added benefit this approach will also prevent transpiration because tree, plant, and shrub roots would be prevented from crossing the barrier. In some cases, the

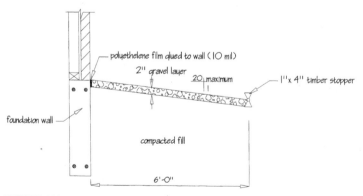

FIGURE 10.2 Horizontal moisture barrier consisting of a polyethylene membrane overlain by a thin gravel layer (from Chen, 1988).

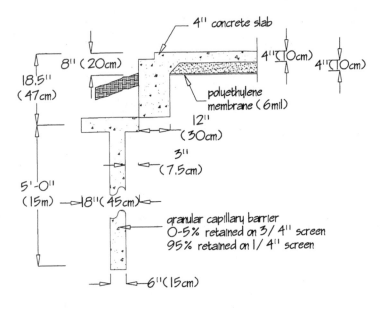

4" concrete slab

8" (20cm)

18.5" (47cm)

4" (10cm) 4" (10cm)

polyethylene membrane (6 mil)

12" (30cm)

3" (7.5cm)

5'-0" (15m) — 18"(45cm)

granular capillary barrier
0-5% retained on 3/4" screen
95% retained on 1/4" screen

6"(15cm)

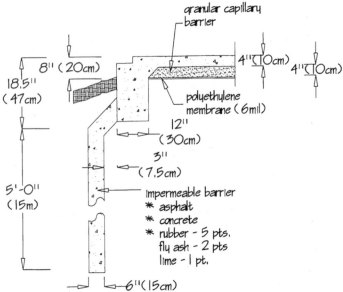

granular capillary barrier

8" (20cm)

18.5" (47cm)

4" (10cm) 4" (10cm)

polyethylene membrane (6 mil)

12" (30cm)

3" (7.5cm)

5'-0" (15m)

impermeable barrier
* asphalt
* concrete
* rubber - 5 pts.
  fly ash - 2 pts
  lime - 1 pt.

6"(15cm)

FIGURE 10.3 Typical permeable and impermeable vertical moisture barriers.

soil moisture within the intended confines of the VICB (prewetting) is increased to a percentage or so above the soil's plastic limit prior to construction of the foundation. (This practice is generally exclusive to slab foundations.) The theory is that the soil is preswelled to a point that increased water is not likely to cause intolerable swell, and at the same time, the barrier should prevent a decrease in soil moisture beneath the foundation. Hence, a stable condition may be created.

Vertical *permeable* capillary barriers (VPCBs) generally consist of a slit trench filled with a permeable material. The VPCB accepts water and distributes it into the permeable barrier. This also tends to block lateral capillary movement of water into the soil matrix itself. (Clay is far less permeable than the material within the barrier.) The use of vertical capillary barriers has not had appreciable success.[15–17,26,58,79]

10.3.3 Conclusion

The use of water or capillary barriers offers possible benefit, but field data made public to date leaves much to be desired.[15–17,26,58,79] For example Chen states, "It is doubtful that the installation (of vertical capillary barriers) is of sufficient benefit to warrant the costs." The installation of moisture controls (such as the French drain, vertical or horizontal moisture barriers), by intent, alters the moisture profile within the foundation-bearing soils. The time period over which the results of this change becomes noticeable might vary from several days to several years. The amount and rate of moisture variation, the particular soil properties, and the foundation design each influence the extent of soil volume change and ultimately any foundation movement.

The installation of French drains is frequently followed by "drying" of the foundation-bearing soils. This could ultimately result in a soil moisture regain if extraneous water becomes available. Unless this gain is uniform over the entire foundation area (which is usually true over the long term), some soil swell and resultant foundation upheaval could eventually occur.

As a matter of interest, the *horizontal permeability* (which translates to lateral water migration) in a highly expansive clay varies from something like 1 ft per year (10^6 cm/s) to 20 ft per year (2×10^5 m/s). (The *vertical* permeabilities are roughly 0.1 ft per year, or 10^7 cm/s). For sands the differences between horizontal and vertical permeabilities are much less, with the general readings being in the range of 1000 to 10 ft per year (10^3 to 10^5 m/s).[65]

10.4 VEGETATION

Certain trees, such as the weeping willow, oak, cottonwood, and mesquite, have extensive shallow root systems that remove water from the soil. These plants can cause foundation (and sewer) problems even if located some distance from the structure. Many other plants and trees can cause different foundation problems if planted too close to the foundation. Plants with large, shallow root systems can grow under a shallow foundation and, as roots grow in diameter, produce an *upheaval* in the foundation beam. Construction most susceptible to this include flat work such as sidewalks, driveways, and patios as well as some pier and beam foundations. Pruning the trees and plants will limit the root development. Watering, as discussed earlier, will also help.

Plants and trees can also remove water from the foundation soil (transpiration) and cause a drying effect that in turn can produce foundation *settlement*. [The FHA (now HUD) suggests that trees be planted no closer than their ultimate height. (There is no basis in fact that relates the lateral spread of roots to tree height.) In older properties this is often not feasible because the trees already exist. With proper care, the adverse effects to the foundation can be minimized or circumvented.]

The principal moisture loss that would be likely to affect foundation stability occurs generally between the field capacity and the level of plant wilt[16,17,69] (see Fig. 8.2). Don Smith, a professor of botany at the University of North Texas, believes that tree roots or other plant roots are not likely to grow beneath most foundations. This is due to several factors, the most important of which are

1. Feeder roots tend to grow laterally within the top 24 in (0.6 m).[49,50,98] The perimeter beam often extends to near that depth and would block root intrusion.

2. Roots prefer loosely compacted soil (low overburden).

3. Soil moisture (long range) and oxygen availability, both necessary for plant growth, are less abundant beneath the foundation.

4. These confined and sheltered areas have no normal access to a replenishing source for water. (Roots tend to "grow to water.")

For the foregoing reasons, it would appear that trees pose no real threat to foundation stability other than that noted in the first paragraph. Along these same lines, even in a semiarid area with highly expansive soil, it is seldom that a significant earth crack is noted beneath the tree canopy. This is particularly true with trees exhibiting low canopies. This suggests an actual conservation of soil moisture.[107] Also, if trees pose the problems that some seem to believe, why don't *all* foundations with like trees in close proximity show the same relative distress? In literally thousands of instances where foundation repair is made without removal of trees, why doesn't the problem at least sometimes recur? Figure 10.4 depicts an actual condition where trees are growing in close proximity to a slab foundation without mishap. This figure shows a real-world representation of the influence of trees on foundations. The tree shown is a pin oak that was planted at the time of construction. The tree is 17 in (43 cm) in diameter, approximately 36 ft (11 m) in height, and has a

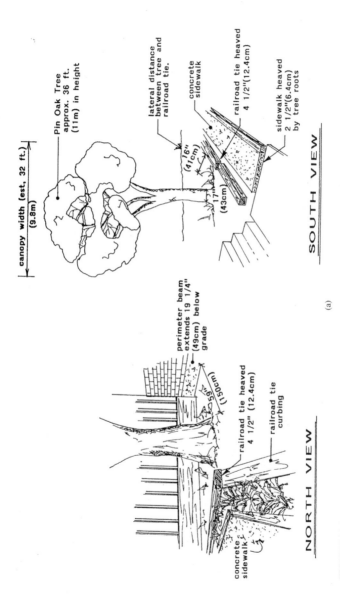

canopy width (est, 32 ft.)
(9.8m)

Pin Oak Tree
approx. 36 ft.
(11m) in height

lateral distance
between tree and
railroad tie.

16"
(41cm)

17"
(43cm)

concrete
sidewalk

railroad tie heaved
4 1/2"(12.4cm)

sidewalk heaved
2 1/2"(6.4cm)
by tree roots

SOUTH VIEW

(a)

perimeter beam
extends 19 1/4"
(49cm) below
grade

59"
(150cm)

railroad tie heaved
4 1/2" (12.4cm)

railroad tie
curbing

concrete
sidewalk

NORTH VIEW

FIGURE 10.4 Tree in close proximity to foundation with no effect on foundation. (*a*) Site drawing.

(b)

FIGURE 10.4 (*Continued*) Tree in close proximity to foundation with no effect on foundation. (*b*) Photograph looking north; note heave of railroad tie, left center.

canopy width of about 32 ft (9.8 m). The tree is located 59 in (150 cm) from the perimeter beam, which extends 15¼ in (49 cm) below grade. There are four pin oaks similarly planted along the west perimeter. The depth of the beam probably accounts for the lack of impact upon the foundation.

Removal of existing trees can create more problems than might originally exist.[15–17,26,42] Cutting (trimming) the roots can induce similar problems, although to a lesser extent.[53] Any extended differential in soil moisture can produce a corresponding movement in the foundation. If the differential movement is extensive, foundation failure is likely to result.

Even with proper care, foundation problems can develop. However, consideration and implementation of the foregoing procedures will afford a large measure of protection. It is possible that adherence to proper maintenance could eliminate perhaps 40 percent of all serious foundation problems. Anyone who can grow a flower bed can handle the maintenance requirements. Remember, one of the basic laws of physics is that nothing moves unless forced to do so. The foundation is no different. This has been emphasized in prior chapters. All foundation repair accomplishes is to restore the structural appearance. If the initial cause of the problem is not identified and eliminated, the problem is likely to recur.

(c)

FIGURE 10.4 (*Continued*) Tree in close proximity to foundation with no effect on foundation. (*c*) Photograph looking south; note heave of sidewalk and north end of railroad tie (in the grass).

CHAPTER 11
FOUNDATION INSPECTION AND PROPERTY EVALUATION FOR THE RESIDENTIAL BUYER

11.1 INTRODUCTION

If you are house shopping in a geographic area that has a known propensity for differential foundation movement, it is always wise to engage the assistance of a qualified foundation inspection service. This could involve an experienced foundation repair contractor or an equally qualified professional engineer (P.E.). The word *qualified* cannot be overemphasized. It is necessary to evaluate the existence of foundation-related problems, determine the cause of the problems, and, when such are found to exist, offer a repair procedure. The cause is particularly important. *Repairs will be futile if the original cause of the distress is not recognized and eliminated.* The principal problem herein lies with the separation of settlement from upheaval (see Chaps. 2, 3, and 5). The checklist in Table 11.1 presents several of the more obvious evaluations of distress relative to differential foundation movement. With only limited experience, one can learn to detect these signs. The problems of detection become more difficult when cosmetic attempts have been used to conceal the evidence. These activities commonly involve painting,

TABLE 11.1 Inspection Checklist

1. *The extent of vertical and lateral deflection.* Does any structural threat exist or appear imminent? Most authorities seem to accept a maximum differential movement of 0.3 percent (over 10 ft this is 0.36 in; over 20 ft this is 0.72 in) as tolerable for normal residential (frame) construction. Table 11.2 gives typical values for various angular distortion (δ) utilizing the "three-point method."[7,73] In the real world of remedial concern, a two-point (δ)' test is normally the procedure utilized. This is due, in large part, to partition walls and other obstructions that prevent or hinder the three-point method. The cited 0.3 percent equates to 1/333, or 1 in/27.75 ft. The wall height has a direct bearing on crack width resulting from δ.[7] At the identical δ (or δ') the crack width in a 16-ft wall will be twice that observed in an 8-ft wall, all other factors being equal.

2. *Whether the stress is ongoing or arrested.*

3. *The age of the property.*

4. *The likelihood that the initiation of adequate maintenance would arrest continued movement.* For example, (1) in cases of upheaval, elimination of the source for water will often arrest movement, and (2) where minor settlement is involved, proper watering may reverse or eliminate the problem.

5. *Value of the property as compared to repair costs.* Most foundation repair procedures require some degree of compromise. To arrive at a reasonable or practical cost, the usual primary concern is to render the foundation "stable" and the appearance "tolerable." In all cases the primary goal should be a "cost-effective" solution. In virtually all repair, the cost to truly level a foundation, if it were possible, would be prohibitive. This is further complicated by the fact that foundations are not generally built level.[15–17]

6. *Type and condition of the existing foundation.* If the foundation is pier and beam, is there adequate crawl space? If it is a slab, was it poured with proper thickness, beams, and reinforcing steel?

7. The possibility that, if the movement appears arrested, cosmetic approaches would produce an acceptable appearance.

patching, tuck-pointing, addition of trim, installation of wall cover, etc. The important issue in all cases is to decide whether the degree of distress is sufficient to demand foundation repair (see also Chap. 9). This decision requires extensive experience and some degree of compromise. Several factors influence the judgment.

A simple way to monitor future movement is to place a pencil mark on various doors, sheetrock cracks, and mortar separations as illustrated by Fig. 11.1. Use a straightedge and sharp pencil. On cracks place the marks such that horizontal as well as vertical displacements are monitored. Initial, date, and where applicable, record the width of the crack. Repeat this process at random locations throughout the property. Future inspections will

TABLE 11.2 Angular Distortion, δ/L.

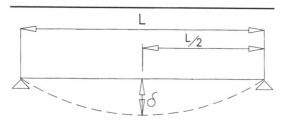

Two Points: $\delta' = \dfrac{\delta}{L/2} = \dfrac{2\delta}{L}$

L, ft (in)	δ/L, %*	δ'/10 ft, in†	δ'/20 ft, in
12.5 (150)	0.66	0.8	1.6
20.8 (250)	0.4	0.48	0.96
25 (300)	0.33	0.4	0.8
	0.31	0.375 (3/8)	0.6875 (11/16)
33 (400)	0.25	0.3	0.6
42 (500)	0.2	0.24	0.48
50 (600)	0.167	0.2	0.4

*δ is constant at 1 in and L varies as shown.
†δ' is distance indicated (120 or 240 in) times 2 δ/L.
Source: After U.S. Dept. of Navy, 1988, and Whitman and Lambe, 1979.

provide the answers to the ongoing movement question. If any movement occurs, the marks will no longer line up. The measured crack widths previously recorded will also show movement. This method not only shows movement but provides a guide as to direction in which movement occurs.

It is difficult, if not impossible, to properly evaluate the above without extensive firsthand experience with actual foundation repairs. To quote Terzaghi, the founder of soil mechanics, "In our field (engineering and geology) theoretical reasoning alone does not suffice to solve the problems which we are called upon to tackle. As a matter of fact it (engineering and geology) can even be misleading unless every drop of it is diluted by a pint of intelligently digested experience."[48] One good example of the importance of on-the-job experience is the proper determination of upheaval as opposed to settlement. If upheaval were evaluated as settlement, the existence of water beneath the foundation would probably be overlooked. In that case, future, more serious consequences would be nearly certain. However, there are usually no similar consequences in making the error of labeling settlement as upheaval. Essentially, this is true because first, all foundation repair techniques

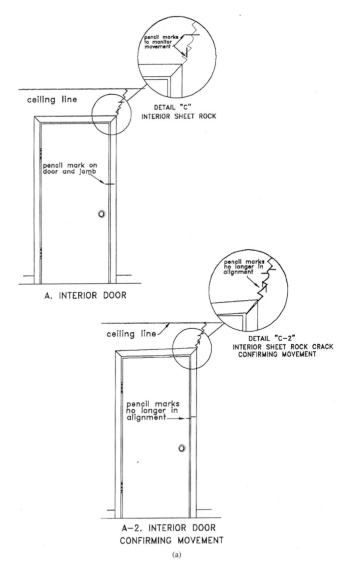

FIGURE 11.1 Monitoring ongoing movement.

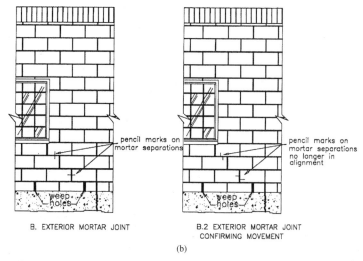

B. EXTERIOR MORTAR JOINT   B.2 EXTERIOR MORTAR JOINT
CONFIRMING MOVEMENT

(b)

FIGURE 11.1 *(Continued)* Monitoring ongoing movement.

are designed to raise a lowermost area to some higher evaluation, and second, there is no undisclosed cause for continual problems. In cases of settlement, one merely raises the distressed area to the as-built level. In instances of upheaval, it becomes necessary to raise the as-built level to approach the elevation of the distorted area. The latter is obviously far more difficult. The existence of foundation problems need not be particularly distressing so long as the buyer is aware of potential problems before the purchase. As a rule, the costs for foundation repair are not relatively excessive, the results are most often satisfactory, and the foundation is stronger and more nearly stable.[3] It is the surprise that a buyer cannot afford.

11.2 FOUNDATION INSPECTION AND SELECTING AN INSPECTOR

The cost for a residential foundation inspection service is quite nominal, varying from about $150 to $500 (1998 dollars), depending on the time involved and the locale. A word of caution: Be selective in choosing your inspector. Since the late 1980s many institutions such as lenders, insurers, and sophisticated buyers have begun the practice of requiring a structural inspection performed by a registered professional engineer. The engineer should be registered as a structural or civil engineer. In some states (Texas, for example) an engineer registered in any discipline (mechanical, electrical, petroleum, aeronautical, industrial, etc.) can claim to be qualified to make the inspection. More often than not, these people are not qualified by either experience or education. The engineer of choice should be independent, unbiased, and not associated (directly or indirectly) with any self-serving entity—foundation repair company, builder, insurer, etc. If possible, the engineer should have extensive hands-on experience in foundation repair. If not, specific repair proposals should be requested from qualified repair companies. If the engineer accepts (or rejects) any bids proposed by qualified contractors, he or she should do so based on a comprehensive, conclusive mathematical analysis. This avoids both overkill and ensures a competent repair.

Experience has shown that often engineers not properly qualified or schooled in foundation problems tend to present defective or slanted reports. As an example, on separate occasions, the same "engineers" have specified *either* (1) 12-in (30-cm) diameter piers, 6 ft (1.8 m) on centers, and drilled to 10 to 12 ft (3 to 3.6 m) with caged four #5s and sometimes even belled to 24 in (60 cm) *or* (2) an 8-in (20-cm)-diameter straight shaft with three #3s, 6 ft (1.8 m) on centers and drilled to 10 to 12 ft (3 to 3.6 m). The 12 in-diameter option contains specifications that (compared to the 8 in-diameter option) are either a gross overkill or apparently intended to remove the 12 in-(30-cm)-diameter pier from price contention. (The ridiculous specifications increase the comparative cost for a 12-in (0.3-m)-diameter pier by about $75.) A 12-in (0.3-m) pier to 10-ft depth on 8-ft (2.4-m) centers reinforced with three #3s would be a superior choice. In this specific example, none of the engineers was registered as either structural or civil. One was registered as a mining engineer and two others as mechanical engineers. This is offered as an observation, not a condemnation, and it does not suggest that errors are not made by well-meaning civil or structural engineers. This happens frequently, usually because of lack of experience. Errors in judgment, intentional or otherwise, tend to produce gross overcharge and/or ineffective repairs.

The engineer of choice should be unbiased and independent as well as competent. It is sometimes difficult for engineers (or anyone else for that matter) to be objective when a decision is contrary to the interest of the party paying the bill. This is particularly evident in the number of engineers who accept or recommend practices, procedures, or causes that have obvious technical flaws.[15–17] Many of these people are either on the payroll of foundation repair companies that "push" a particular process or retained by insurance companies with self-serving agendas.

Evaluate any disclaimers included in the engineer's report. If the engineer specifies the method of repair, he or she probably assumes legal responsibility for the outcome of the repair and future stability of the structure. This is probably true, despite any disclaimer included in the report. These disclaimers, often used as some attempt to shield the engineer from inexperience, probably would not survive an appropriate court of law. The repair contractor would probably be named as a third-party defendant in any litigation resulting from repair; however, the ultimate responsibility would probably fall on the engineer who designed and specified the repairs. In this event, the consumer would be lucky if the defendant engineer carried insurance. Otherwise, consumers are apt to be without recovery unless their state requires bond coverage for practicing engineers. This situation can be avoided (somewhat at least) by the engineer (or owner) soliciting proposals from qualified foundation repair companies who specify and, in turn, provide warranty for all work. The question now becomes How can one determine the presence of a potential foundation problem?

11.3 INDICATIONS OF STRUCTURAL DISTRESS

Following is a simple checklist for evaluating the stability of a foundation. If you have questions or uncertainty regarding any of the items, consult a qualified authority. One might also refer to Sec. 9.4 to evaluate the "severity" of any observation. Figure 11.2 presents photographs showing several of the following concerns:

1. Check the exterior foundation and masonry surfaces for cracks, evidence of patching, irregularities in siding lines or brick mortar joints, separation of brick veneer from window and door frames, trim added along door jam or window frames, separation or gaps in cornice trim, spliced (extended) trim, separation of brick from frieze or fascia trim (look for original paint lines on brick), separation of chimney from outside wall, masonry fireplace distress on interior surfaces, etc.

2. Sight ridge rafter, roof line, and eaves for irregularities.

3. Check interior doors for fit and operation. Check for evidence of prior repairs and adjustment such as shims behind hinges, latches or keepers relocated, tops of doors shaved.

4. Check the plumb and square of door and window frames. Are the doors square in the frames? Check to see if the strike plates have been adjusted to accommodate the strikers. A relocation might indicate movement. Measure the length of the door at the door knob side and the hinge side. A discrepancy suggests that the door may have been shaved. Also feel the top of the door above the door knob. If it is smooth, the door may have been sanded or shaved. If the door rubs slightly at the top, shimming the hinge plate might provide alignment without altering the door.

5. Note grade of floors. A simple method for checking the level of a floor (without carpet), window sill, countertop, etc., is to place a marble or small ball bearing on the surface and observe its behavior. A rolling action indicates a "downhill" grade. (A hard surface such as a board or book placed on the floor will allow the test to be made on carpet).

(a)

(b)

(c)

FIGURE 11.2 Evidence of differential foundation movement. (*a*) Brick veneer tilted inward; often an indication of interior slab settlement. (*b*) Separation of brick veneer from garage door jamb. (*c*) Brick veneer separated from window frame.

(d)

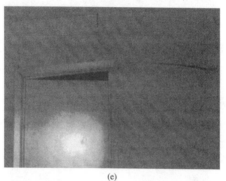

(e)

(f)

FIGURE 11.2 *(Continued)* Evidence of differential foundation movement. (*d*) Horizontal separation in brick mortar. (*e*) Interior door frame out of plumb causing sheetrock cracks. This illustrates the effect of upheaval in an interior slab. The high point is to the right side of the door. (*f*) Interior slab settlement. This shows separation of interior slab from the wall partition. Note also that the bathtub has settled away from the ceramic tile.

11.9

(g)

(h)

(i)

FIGURE 11.2 *(Continued)* Evidence of differential foundation movement. (*g*) Crack in perimeter beam. This illustrates a major crack in the perimeter beam accompanied by secondary cracks in brick veneer and brick mortar. (*h*) Stair step separation in brick mortar. This shows an obvious crack in the brick mortar, but does it represent a problem of concern? First, the crack width is less than $\frac{1}{4}$ in (0.6 cm). Second, the mortar joints are straight. Third, the cornice trim at the corner is tight. Fourth, no interior damage was noted. Thus, the damage is not one of concern. (*i*) Slab upheaval. A fairly typical representation of slab upheaval. Note the very obvious crown in the slab surface (approximately 4 in, or 10 cm) accompanied by vertical separation in wallboard, cracks in floor slab, and separation of shoe mold from slab.

6. Inspect wall and ceiling surfaces for cracks or evidence of patching. *Note:* Any cracking should be evaluated on the basis of both extent and cause. Most hard construction surfaces tend to crack. Often this can be the result of thermal or moisture changes and not foundation movement. However, if the cracks approach or exceed $\frac{1}{4}$ in (0.6 cm) in width, the problem is *possibly* structural. On the other hand, if a crack noticed is, for example, $\frac{1}{8}$ in (0.3

cm) wide, is it a sign of impending problems? A simple check to determine if a crack is "growing" is to scribe a pencil mark at the apex of the existing crack and using a straightedge, make two marks along the crack, one horizontal and one vertical. If the crack changes even slightly, one or more of the marks will no longer match in a straight line along the crack and/or the crack will extend past the apex mark. A slight variation of this technique is to mark a straightline across a door and matching door frame. If any movement occurs, the marks will no longer line up. In a structure older than about 12 months, continued growth of the crack or displacement of the doormarks would be a strong indication of foundation movement.

7. On pier-and-beam foundations, check floors for firmness, inspect the crawl space for evidence of moisture and deficient or deteriorated framing or support, and ascertain if there is adequate ventilation. The crawl space should be dry and adequate for access. As a rule of thumb, 1 ft^2 (0.9 m^2) of vent is suggested for each 150 ft^2 (13.5 m^2) of floor space.

8. Check interior drainage adjacent to foundation beams. Any surface water should quickly drain away from the foundation and not pond or pool within 8 to 10 ft (2.4 to 3 m). Give attention to planter boxes, flower-bed curbing, and downspouts on gutter systems.

9. Look for trees that might be located too close to the foundation. Some authorities feel that the safe planting distance from the foundation is 1 or preferably 1.5 times the anticipated ultimate height of the tree. More correctly, the distance of concern should be perhaps 1 to 1.5 times the canopy width. Consideration should be given to the type of tree. Also, remember that the detrimental influence of roots on foundation behavior is grossly overrated (see Chaps. 3 and 8).

10. Are exposed concrete surfaces cracked? Hairline cracks can be expected in areas that have expansive soils. However, larger cracks approaching or exceeding $1/4$ in (0.6 cm) in width warrant closer consideration.[17,61]

11.4 IMPACT OF FOUNDATION REPAIR ON PROPERTY EVALUATION

Do foundation repairs affect resale values? This is a question often heard, particularly from appraisers and attorneys. Generally speaking competent foundation repairs should produce an end product at least equivalent, and more often, superior, to the original. Figure 11.3 shows before and after photos of actual foundation repair. In all instances, the damage was completely reversed. (This is generally but not always the case.) The repaired foundation will have no increased susceptibility to future problems. In fact, the properly repaired foundation might be considerably more resistant to future distress. These facts are due, at least in part, to such factors as

(a)

(b)

FIGURE 11.3 Foundation leveling. (*a*) The floor is separated from the wall partition by about 4 in (10 cm). (*b*) The separation in brick mortar is in excess of 2 in (5 cm).

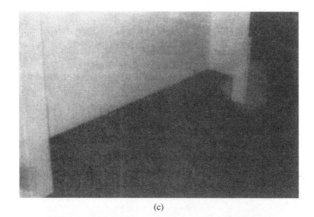

(c)

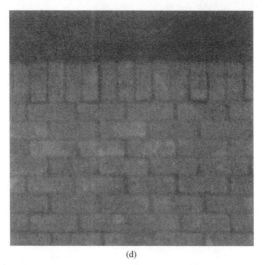

(d)

FIGURE 11.3 *(Continued)* Foundation leveling. (*c*) and (*d*) Results from foundation levelings of (*a*) and (*b*), respectively. The separations are closed. The end results of foundation leveling are not always so impressive.

(e)

(f)

FIGURE 11.3   *(Continued)* Foundation leveling. (*e*) Mudjacking in progress. The shovel and scribed marks monitor the raise of the slab. At this point, the raise has been more than the width of a brick. (3½ in, or 9 cm). This is shown by the distance between the shovel point and the original scribed mark. (The shovel *handle* is on a stationary surface.) (*f*) In this example the shovel *point* is on the stationary surface.) The floor slab is down about 4 in (10 cm).

(g)

FIGURE 11.3 *(Continued)* Foundation leveling. (*g*) The floor is restored to its original grade.

1. Underpinning and mudjacking provide support to the foundation that either did not preexist or had been rendered ineffective.

2. The cause of the foundation failure was (it is hoped) identified and corrected prior to the repair. [With the new (unrepaired) foundation some problem might exist but will not be detected until a foundation problem actually develops.]

3. Certain repair procedures (such as mudjacking or chemical stabilization) actually help stabilize the soil moisture, inhibit moisture transfer within the soil, and increase soil resistance to shear. (The grout placed by mudjacking tends to serve somewhat as horizontal and vertical capillary barriers.)

4. A foundation properly repaired by a reputable contractor will be covered by a warranty of some duration. The *legitimate* warranties generally vary in length from 1 to 10 years with restrictions or limitations in coverage after the first year or so. Read any so-called lifetime or extended-term warranty carefully. If it sounds too good to be true, you can be assured *it is.*

A foundation is no different from any other inanimate object—it refuses to move unless forced to do so. Eliminate or prevent this force and no movement occurs. In the case of foundations built on expansive soils, the primary force is provided by variations in soil moisture. Add water and the soil swells. Remove water and the soil shrinks. Do neither and no movement occurs. It makes no difference whether the foundation is new, old, or previously repaired.

Proper maintenance and alert observation to the early warning signs of impending distress will prevent or minimize all foundation movement.

Drainage and proper water can be easily controlled by common sense. The "unknown" in the formula involves the accumulation of water beneath slab foundations due to some form of utility leak. The most serious source is some form of sewer leak. As a rule the latter is usually detected after the fact from the various signs of differential foundation movement manifested by sheetrock cracks, ill-fitted doors, windows, cabinets, etc., or deviated floors. Figure 9.3a shows a simplistic slab behavior as it is being heaved. Figure 9.3b shows what happens to the rebar after the slab is heaved. Catch the problem sufficiently early, eliminate the cause, and in some cases, avoid the need for foundation repair. Notice that no differentiation is given as to age or history of the subject foundation. It makes little difference.

The principal foundation concern in purchasing a property should be whether the foundation was properly designed for the site conditions, competently constructed, and properly maintained. The occurrence of foundation repair would have no negative impact, providing the repairs were properly executed.

11.5   WARRANTIES AND SELECTING A REPAIR CONTRACTOR

Once the method of repair has been resolved, the selection of a contractor has been narrowed considerably. A next point to consider might be the contractor's warranty.

11.5.1   Warranties

A foundation will not move unless forced to do so. The force causing foundation movement (on expansive soils) is generally water—too much or too little. The contractor has no control over the availability of water to the foundation. Therefore most contractor's warranties limit liability for upheaval, whether caused by deficient drainage (ponding), excessive watering, or domestic sources. Excluding these events, the foundation is not likely to experience significant movement anyway. The quality contractor's *standard* limited warranty (usually 1 to 2 years) will frequently (1) waive the upheaval restriction on a limited basis, (2) provide transfer to new owners, and (3) cover differential movement in excess of $1/4$ to $3/8$ in (0.6 to 0.9 cm). [Lower ranges of movement are allowed because virtually all foundations on expansive soils will move. In fact, slab foundations were originally designed to accommodate soil movement (often 0.3 percent), hence, the term *floating slab*. Refer to the Introduction and Sec. 9.4.] Many contractors also offer a limited *extended* warranty to cover the repaired foundation for an extended period of time, usually 3 to 10 years. Repairs under the extended coverage would cost the consumer a reduced amount, often about one-half normal charges. The extended coverage is frequently free of extra cost to the consumer, unless, of course, work is actually performed.

Be wary of lifetime warranties or other ridiculous warranty claims. The verbiage of the lifetime warranty often includes several or all of the following caveats:

1. Covers *initial owner only* or reduces to 10-year warranty if property changes hands within 10 years.

2. Does *not cover* foundation movement due to *heave*. (Over 70 percent of all slab foundation repairs are precipitated by upheaval.)

3. Specifically excludes settlement (or movement) of interior floors.

4. Covers materials only.

5. Limits coverage to replace or rework *material* installed by contractor. Additional material or "new" installations are subject to normal charge.

6. Limits coverage to *vertical settlement*.

7. Requires differential deflection of 0.4 percent (1 in over 240 in) as opposed to the "normal" 0.3 percent (1 in over 360 in) to define failure. Refer to Sec. 9.4.

8. Exempt all "consequential" damage such as landscaping, floor covering, sheetrock, brick, masonry or concrete, utilities, etc. (This is basic to most, if not all, warranties.)

The terms *lifetime* and *warranty* are somewhat deceptive, to say the least. Aside from the foregoing, for a lifetime warranty to be taken seriously it would need to be backed by a *substantial* irrevocable, escrow account. The balance for such a fund should start at something like $500,000 and be increased annually as the company increases exposure. A half million dollars would assure full dollar coverage on a mere 30 to 100 jobs, dependent, of course, on the size of the composite projects and assuming that neither punitive nor treble damages become involved. (A major foundation repair company might perform upward of 40 jobs per month.) Many warranties, particularly the so-called lifetime ones, are reflective of insuring swimming pools against theft. The rates are quite cheap and the coverage can be lifetime.

When all is said and done, the best and most consumer protective warranty is probably represented by the "normal" 2-year standard limited warranty. In effect, these warranties state "Contractor warrants work performed by the contractor against *settlement* in excess of $1/4$ inch for a period of 24 months after completion of initial work." Note particularly the reference to settlement. Most warranties (and contractors) exclude upheaval because this problem is caused by water accumulation and is beyond the control of the contractor.

Most recurrent problems, which are generally the responsibility of the contractor, occur within the first year. Even then the problems are likely to be quite minor. The principal exception to this is those circumstances where (1) a slab foundation is underpinned without the prerequisite mudjacking or (2) defective methods are used to underpin the perimeter and proper mudjacking is not performed. Sometimes, more than 2 years are required for these problems to become apparent to the consumer. However, the extended limited warranty (5 to 10 years) will adequately cover any of these contingencies.

Upheaval represents the largest cause for recurrent foundation failure of slab foundation on expansive soils. This problem is exempted from most, if not all, warranties. However, some companies will accept this as a covered item *strictly* as a public relations concession. Where the concession is made, the coverage is provided by the standard limited warranty.

The foregoing analysis is understood when one remembers the principal of physics—nothing moves unless *forced* to do so. With slab foundations or expansive soils, this force is, generally, water and, most often, excess water. This instance occurs beyond the control of even the most conscientious contractor.

In other cases, recurrent problems can result that are not related to expansive soil behavior. Generally, these are also exempted by the contractor and might include such instances as sliding or embankment failure, erosion, consolidation of fill due to decay of organic content or collapse of soil structure, thaw of permafrost, etc. (see Chap. 8).

If you have a warranty concern, consult your attorney for an opinion. Be more concerned over what the warranty covers and less about the duration.

11.5.2 Selecting a Contractor

Next in the process to find the preferred contractor is to consider individual credentials. To help ensure the selection of a competent contractor, consider the following criteria:

1. The contractor has *at least* 5 years of experience (preferably 10) in your locale dealing with problems similar to your own. Request proof of experience. Contact the Yellow Pages to determine the contractor's continuous years therein. There are companies in the Yellow Pages who advertise X years experience. In reality, a number of different employees each have a few years experience, which totals X.

2. The contractor should be fully insured and licensed, where applicable.

3. The contractor should be financially stable. A bankrupt contractor offers little or no protection or recourse for complaints.

4. Check with the Better Business Bureau, Attorney General, and County and District Courts to evaluate litigation and/or unresolved disputes. Any contractor in business for an extended period of time is subject to complaints. The concern is how the disputes were resolved.

5. Does the contractor have adequate technical competency? Repairs are neither repetitious nor sometimes even similar. The contractor must have the ability to adapt.

6. Ascertain that the contractor owns all the proper equipment with which to do your job. Use of unfamiliar equipment has been known to result in serious property damage. The contractor should do the work with company personnel. In some cases, *limited* subcontracting has proven acceptable.

With careful reliance on the foregoing paragraphs, an honest, capable, and economical foundation repair should result. As stated in Chap 8, the principal goal is to identify and eliminate the cause of the problem. This done (and assuming competent repairs), recurrent movement is unlikely unless a new cause develops.

11.6 BUILDING CODES

New construction is governed, to a large extent, by municipal or national codes.[29,36,90,93] Most major cities have written codes allegedly based on the Council of American Building Officials (CABO) publication for one- and two-family dwellings, 1989 ed. This information does not provide any real assistance, as far as foundation repair is concerned. In fact, the author is not aware of any significant industrial, municipal, national, or governmental code covering foundation repair.

Many municipalities in the United States have established city policies intended to control at least certain aspects of foundation repair. Generally, this is through the requirement of building permits. At the end of the day, about the only benefit from this policy is to the city coffers, at the expense of the citizens. The municipality offers no guidelines as to proper repair procedures. The code merely requires the seal of a P.E. In Texas, the P.E. might be schooled in such areas as aeronautical, sanitary, industrial, electrical, mechanical, agricultural, petroleum mining, or if the citizen is extremely lucky, civil or structural engineering. In many cases, this "engineer" has never leveled a foundation and does not provide any mathematical analysis that would verify the design or support his or her credibility.

11.7 CONCLUSIONS

Regardless of the foundation repair contractor's skills or methods, he or she can generally locate an engineer who is willing to sign off on any proposal and/or drawing(s). This secures a building permit. Next, the city requires inspections with the intent of ascertaining that the work performed is according to the engineer's approved design—not with any regard to the *merit* of the repairs. Generally, city inspections provide little service to the consumer, unless a less than honest contractor has been hired. The city inspectors can, sometimes, threaten the success of a job, principally by delaying the pouring of concrete. Drilled pier shafts should be poured as quickly after drilling as possible. An overnight delay can be detrimental to the function of the drilled pier/haunch.

A reasonable and effective code covering foundation repair would be a substantial asset to the owners and repair contractors alike. However, *serious* thought must be given to establish a valid, workable, reasonable code (see Chaps. 2 through 7). So much "hocus pocus" is written about foundation failure and repair that it has become increasingly imperative to separate fiction from fact. Writers have attempted to base a *trend* on one or two questionable case histories. In other cases, repair procedures have not been subjected to a *thorough* mathematical analysis to evaluate such characteristics as component resistance to (1) shear, tensile, lateral, or compressive stress; (2) moments that induce misalignment or cause failure; (3) potential heave; (4) chemical attack or deterioration; and (5) load attributed to the weight of the structure.

The specific design of the foundation, the site conditions, and the bearing soil characteristics are examples of other factors to be considered when adopting repair procedures. Proper forethought and the input of knowledgeable, experienced, unbiased contractors in concert with equally capable engineers, geotechs, and/or perhaps architects could well develop a usable foundation repair code. Actually, a *series* of codes applying to variable conditions is needed.

CHAPTER 12
CASE HISTORIES

12.1 INTRODUCTION

There are thousands of potential foundation repair case histories. The following were selected principally because they offered something a little different.

12.2 FLORIDA LAKE HOUSE

This problem involved extreme subsidence brought about principally when the water level in an adjacent lake was lowered several feet. The bearing soil consisted of a top layer of silty sand, a midlayer of decayed organics (peat), and a base of coral sand.

The two-story brick-veneer dwelling suffered from differential foundation settlement of 6 in (15 cm) in magnitude. The objectives were to (1) underpin the perimeter beam to facilitate leveling (as well as provide future stability), (2) consolidate the peat stratum by pressure grouting to provide a solid base for the repaired structure, and (3) mudjack the entire foundation slab area to create a level or more nearly level structure. Figure 12.1 depicts the repair process. First, excavations were made at strategic locations to provide the base for the spreadfootings (underpinning). Next, grout pipes were driven through the base of the excavations into the peat identified for consolidation. Steel-reinforced concrete was then placed in the excavations and allowed to cure. While the footing pads were curing, the entire slab was drilled for mudjacking and interior deep grouting. Deep grouting was then initiated, starting with the permanent grout pipe set through the footings and continuing to the interior.

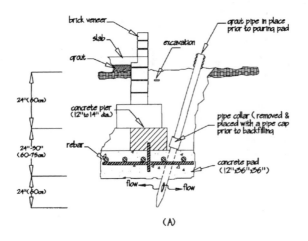

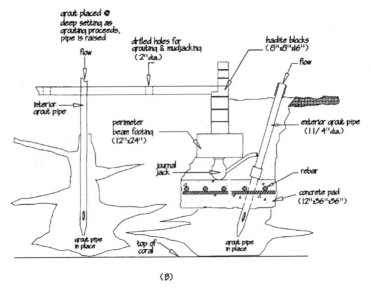

FIGURE 12.1 Florida repair. (*a*) Grout pipe in place; (*b*) typical deep grouting procedure.

At each site, grouting was continued either to refusal, a breakthrough of grout to the surface, or the start of an unwanted raise. Interior grout pipes were removed after the grout process. The permanent, exterior pipes were cleared of grout and capped. After the grout had set, the top section of the perimeter pipes was removed and capped below grade to permit regrouting at some future date should subsidence recur. Next, the perimeter beam was raised to desired grade and pinned by the installation of the poured concrete pier caps. The final step was to mudjack the entire foundation for final grading. Repairs were completely successful and have remained so since 1980. Prior to the work, the dwelling could not be inhabited or sold.

12.3   GARAGES IN LONDON, ENGLAND

The soil in the United Kingdom is not materially different from that found in parts of the United States. The plasticity indices classically run in the range of 40 to 50, and the problem clay (montmorillonite) content is in the vicinity of 20 to 40 percent. The annual rainfall is not particularly high, only about 30 in (76 cm) of rain per year. The rain is well distributed over the year (150 days), with seldom more than 3 in (7.5 cm) during any one month. The Dallas Metroplex, for example, has about 30 in (76 cm) of rain per year but perhaps 75 to 80 percent of the total falls in less than 15 days. (The latter experiences something like 90 percent run-off.) Further, London's high temperatures are generally in the 70°F (21°C) range, whereas the Metroplex highs exceed 105°F (40.5°C). Thus, London's climate produces a high, fairly consistent moisture content within the soil.

Often the soil moisture persistently approaches or exceeds the plastic limit (PL), indicating little, if any, residual swell potential. [It is interesting to note that moisture contents taken from soil borings in close proximity to trees often show little, if any, reduction in percentage of water between the depth of approximately 3.3 ft (1 m) to perhaps 49 ft (15 m). Obviously, this suggests that, over the centuries, the soil has attained a level of unique moisture balance.] Occasionally, a prolonged change in climate does come along that tends to temporarily disturb the balance, such as the drought of 1976. During that period, the soil moisture within the top 6.6 ft (2 m) or so was significantly reduced, reportedly causing severe and extensive problems of subsidence. Later, upon return of the normal moisture, the problems became even more severe due to soil swell and upheaval.

The London projects involved restoring the foundation of four banks of garages to the extent that the repairs could be expected to alleviate future distress for a period of at least 20 years. The repair procedure included the pressure injection of Soil Sta into the bearing soil beneath the foundations to a depth of 6 ft (1.8 m). The chemical was injected on the basis of $1/4$ gal/ft^2 (1 mL/cm^2)

of the surface treated. Soil Sta was used in the hope of precluding the recurrence of the effects of drought conditions such as those of 1976, specifically the upheaval phase. Next, spreadfootings were installed beneath the load-bearing perimeter to permit mechanical raising and underpinning. Soil Sta was injected through the base of each footing excavation prior to pouring concrete. The chemical volume and purpose were the same as specified above. See Figure 1.12 for a detail of the foundation design. The final stage involved mudjacking to fill voids, raising and leveling the slab foundation, and filling any voids beneath the perimeter beam that resulted from the underpinning.

The jobs were considered successful. The procedure was substantially less expensive than deep pilings or needle piers, which were previously considered as a conventional approach. In addition, the foregoing methods caused less damage to the landscaping and were quicker and less involved to perform.

12.4 PIERS—DRIVEN STEEL PIPE: AN EXAMPLE OF FAILURE

The field photographs in Figs. 5.9 and 5.10 show steel pipe placed by a hydraulic driver and pinned by bolts through the lift bracket. Figure 5.9a shows the pipe slanted inward at over 30° (proving nonalignment). In this instance, if one assumes an axial load on each steel pile of 6000 lb (27 kN), the lateral vector (f_x) would be 3000 lb (13.5 kN). This force plus any lateral force created by the soil can be responsible for pile failure in lateral stress (see Sec. 8.4.1 and Fig. 12.2). The lateral component of soil stress is also discussed in a section of the book by Prakash and Sharma[91] and in an article by G. G. Myerhoff.[74]

Figure 5.9b shows the lift bracket not in contact with the beam. (The settlement of the pipe could be the result of soil dilatency, clay-bearing failure, or ultimate failure in whatever material or object into which the pier tip is embedded.) The shiny spots on both pipes represent a prior attempt to adjust the pipes to reraise the beam. Figure 5.9c represents the excavation of two minipiles at the corner of a foundation. Note the obvious bending and loss of contact between the perimeter beam and lift brackets. These piers are totally ineffective. Figure 5.9 depicts only a few examples; however, this type of performance appears to accompany the driven steel minipipe procedure at least when expansive soils are involved. Refer also to Sec. 5.4.

Other conditions of load-pile/pier reaction exist. For example, a vertical pile subjected to an inclined load Q at angle w is equivalent in behavior to a batter (rake) pile/pier inclined at an angle w and subject to vertical load Q. The simplest and preferred condition occurs when the pile/pier is vertical with the load applied concentrically.

It certainly seems safe to say that nonperformance represents the rule rather than the exception, at least within certain areas. Another problem, limited to

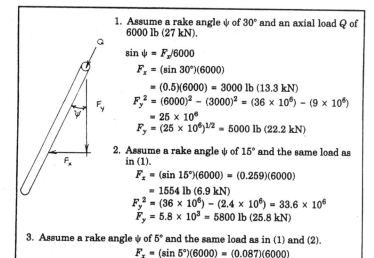

1. Assume a rake angle ψ of 30° and an axial load Q of 6000 lb (27 kN).

$$\sin \psi = F_x/6000$$

$$F_x = (\sin 30°)(6000)$$

$$= (0.5)(6000) = 3000 \text{ lb } (13.3 \text{ kN})$$

$$F_y^2 = (6000)^2 - (3000)^2 = (36 \times 10^6) - (9 \times 10^6)$$

$$= 25 \times 10^6$$

$$F_y = (25 \times 10^6)^{1/2} = 5000 \text{ lb } (22.2 \text{ kN})$$

2. Assume a rake angle ψ of 15° and the same load as in (1).

$$F_x = (\sin 15°)(6000) = (0.259)(6000)$$

$$= 1554 \text{ lb } (6.9 \text{ kN})$$

$$F_y^2 = (36 \times 10^6) - (2.4 \times 10^6) = 33.6 \times 10^6$$

$$F_y = 5.8 \times 10^3 = 5800 \text{ lb } (25.8 \text{ kN})$$

3. Assume a rake angle ψ of 5° and the same load as in (1) and (2).

$$F_x = (\sin 5°)(6000) = (0.087)(6000)$$

$$= 522 \text{ lb } (2.3 \text{ kN})$$

The lateral component of the axial force Q reduces to zero as the rake angle ψ approaches 0°. This is an overly simplified example, but it serves to illustrate a point.

FIGURE 12.2 Calculation of lateral force on raked steel minipile.

slab foundations, has been the failure of contractors to follow the piling process with competent mudjacking of the slab. Because the slab is not designed to be a bridging member, voids, preexistent or created by raising the perimeter, encourage interior settlement of the floors. This must be circumvented by mudjacking. Proper mudjacking could, in fact, eliminate or minimize some of the other inherent deficiencies of the driven-pipe process. Mudjacking alone will normally hold a raised slab foundation, provided proper maintenance procedures are instituted and followed (Chap. 10).

12.5 RAISING AND LOWERING A FOUNDATION

On rare occasions, it becomes desirable to both raise and lower sections of a foundation. Figure 12.3 represents such a situation. As the elevations indicate (Fig. 12.3a), the south two-thirds of the foundation has settled significantly

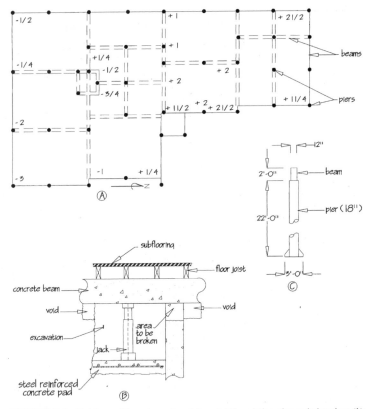

FIGURE 12.3   Raising and lowering a foundation. (*a*) Foundation plan and elevation; (*b*) schematic drawing of mechanical setup for raising foundation; (*c*) pier detail (existing).

[up to 3 in (7.5 cm)], whereas the garage, north, northwest, and north central beams have heaved [at least $2^1/2$ in (6.25 cm)]. (The elevations are considered accurate to the extent that relative grade positions are shown. The question lies in defining the true *differential* movement.) The floors were not in contact with the soil and the beams were originally underlaid with 12-in (0.3-m) void boxes. The void had been lost in many areas due to either beam settlement, filling in by soil, or some combination of both. The void area was to be restored by excavating beneath the beams as required and/or raising the beam.

Conventional methods were used to raise the settled areas of the beam. Next, the existing piers were adjusted and extended to resupport the beam, instead of adding new drilled piers or spreadfootings (see Fig. 12.3). (Spreadfooting pads

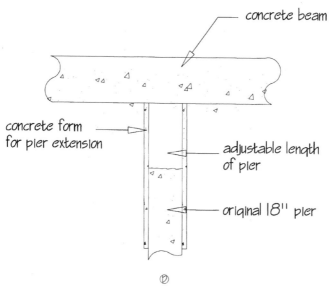

concrete beam

concrete form
for pier extension

adjustable length
of pier

original 18" pier

ⓓ

FIGURE 12.3*d*   Preparing to pour new pier cap extension.

ⓔ

FIGURE 12.3*e*   North wall of garage undermined. Note existence of original piers approximately beneath each window (refer to part *a*—floor plan). The concrete pads to be used in the lowering operation appear at the bottom side of each pier. The pads for each corner pier are not evident.

were used as a base from which to raise the beam sections after the existing belled piers had been broken free from the beams.)

The heaved segments of the beam were lowered by excavating the soil beneath the affected length of the beam, supporting the beam on jacks (resting on spreadfooting pads as previously noted), removing any above-grade lengths of the pier, pouring appropriate additions to the existing pier at each location, reloading the adjusted piers by lowering the beam onto the "new" piers, and then finally removing the jacks.

At the completion of the project, grade elevation suggested that the maximum remaining vertical deflection in the foundation was less than $1/2$ in (1.27 cm), whereas at the beginning the maximum deflection was $5\frac{1}{2}$ in (14 cm). This particular job presents an interesting question. What is the merit of the lowering procedures? In undermining the approximate 250 linear feet (75 m) of beam, it was necessary to remove approximately 300 ft^2 (27 m^2) of floor and subflooring to gain access to the interior beams. This area, plus the perimeter beams, was then excavated (see Figure 12.3e). The cost of undermining, including floor replacement plus the attendant lowering operation, represented about one-half the entire foundation repair bill. This approach was necessary to give the customer what he wanted, but could some compromise have been more cost-effective? Perhaps not in this particular instance because the home was vacant and on the market.

The author tends to generally disfavor lowering operations. Raising lower areas to meet the high is almost always the most practical solution, particularly when dealing with normal residential or light commercial construction. Nonetheless, this procedure does give the repair contractor another option.

12.6 APARTMENT BUILDING

This project involved a more complicated problem. The foundation distress was such that the masonry exterior walls were forced outward to the point where the second-story precast concrete floor slabs were pulled almost off their base. The foundation problem was addressed as a typical slab repair. The perimeter was underpinned (spreadfootings in this case), and the interior floor slab was mudjacked. Concurrently, the interior was crisscrossed with dywidag bars at the first-floor ceiling, extending through the dry wall to the exterior (see Fig. 12.4). The walls were plumbed as tension was applied by tightening the nuts on the dywidag bars against the steel beams or plates. All beams were later replaced with steel plates. A ceiling furr-down was used to conceal the bars on the inside. The exterior bars were cut flush with the nuts, plastered over, and painted to match the exterior walls. (Willard Smith & Associates, Dallas, performed the steel work.)

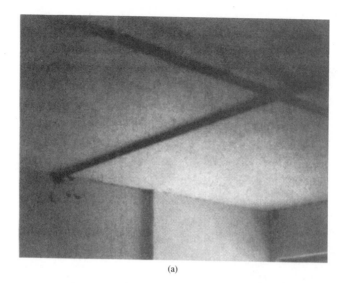

(a)

(b)

FIGURE 12.4 Apartment building repair. (*a*) Dywidag bars crisscross the interior space immediately below the first-floor ceiling; (*b*) steel beams distribute tension and allow walls to be pulled inward by tightening a nut on the dywidag threaded bar.

(c)

FIGURE 12.4 (*Continued*) Apartment building repair. (*c*) Steel beams are replaced with smaller plates once the wall has been plumbed. The excess bar is cut off flush with the nut and the assembly is stuccoed over and painted.

12.7 WACO TEXAS SLAB FOUNDATION NOT PROPERLY MUDJACKED

This represents another case where the slab foundation had been previously repaired and the effort was ineffective. Figure 12.5 shows a photograph of the interior slab at the bath and an artist's concept of the predicament. In this example, the perimeter was underpinned using 12-in (0.3-m)-diameter concrete piers (properly it would seem because the recurrent failure involved only the interior). The interior was supposedly mudjacked. If mudjacking was, in fact, performed, two facts seem apparent: (1) It was not thoroughly or properly completed and (2) there was no evidence of it (patched drill holes) in the area. Restitution involved simply mudjacking the settled areas of the interior slab.

12.8 CONCLUSIONS

Based on the foregoing, it becomes apparent that

1. Foundation repairs tend to principally protect against recurrent settlement.

(a)

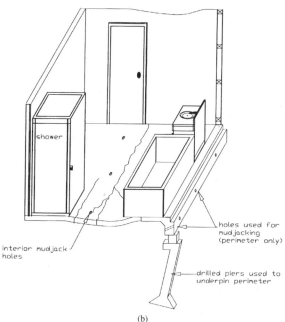

shower

interior mudjack holes

holes used for mudjacking (perimeter only)

drilled piers used to underpin perimeter

(b)

FIGURE 12.5 Interior slab failure due to improper mudjacking. (*a*) Photograph depicting settlement of interior slab. In this instance the condition was caused by incomplete mudjacking during initial repairs. The interior slab was not properly mudjacked. (*b*) Artist's rendering of above condition.

12.11

2. Recurrent upheaval is a concern, especially, with slab foundations, unless there are thorough utility checks and proper maintenance.

3. Spreadfootings appear equal or, in some cases perhaps, superior to "deep piers" as a safeguard against resettlement.[5]

4. "Deep piers" may, in some instances, be conducive to upheaval.[5] Although this observation is certainly true, the occurrence is not frequently documented. [The term *deep piers* is a colloquialism used to refer to piers constructed to a minimum depth of about 10 ft (3 m).[17]]

5. Upheaval accounts for more foundation failures (requiring repair) than settlement by a factor of about 2.3 to 1.0.

6. Moisture changes that influence the foundation occur within relatively shallow depths.

7. The effects of upheaval distress occur more rapidly and to a greater potential extent than do those of settlement.

8. The cause of foundation problems must be diagnosed and eliminated if recurrent distress is to be avoided.

9. Weather influences foundation behavior, both prior to and after construction.

REFERENCES

1. *ACI Reports 302.1R-80, 318-83,* and *435.3R-84,* American Concrete Institute, Detroit.

2. Allen, P. M., and Flanigan, W. D., "Geology of Dallas, Texas, United States of America," Association of Engineering Geologists, 1986. Proceedings of the Seventh International Conference on Expansive Soils, ASCE, Dallas, August, 1992.

3. Baker, T. D., "Acceptance Criteria for Lime Slurry Pressure Injection," Seventh International Conference on Expansive Soils, ASCE, Dallas, 1992.

4. Baker, T. H. W., and Goodrich, L. E., "Heat Pump Chilled Foundations for Buildings and Permafrost," *Geotechnical News,* September 1990.

5. Blacklock, J. R., and Pengelly, A. D., "Soil Treatment for Foundations on Expansive Soils," ASCE Paper, May 1988, Nashville.

6. Bobbitt, D. E., "Theory and Application of Helical Anchors for Underpinning and Tiebacks," Helical Piers and Tieback Seminar, Atlanta, March 1990.

7. Boone, Storer, "Ground Movement—Related Building Damage," *JGE,* November 1996.

8. Borden, R. H., Holz, R. D., and Juran, I., *Grouting, Soil Improvement and Geosynthetics,* Vols. 1 and 2, ASCE, New York, 1992.

9. Bouwer, H., Pyne, R. A. G., and Goodwich, J. A., "Recharging Ground Water," *Civil Engineering,* June 1990.

10. British Standards Institute, *Code of Practice for Foundations,* Report CP2004, 1972, p. 31.

11. Broderieck, G. P., and Daniel, D. E., "Stabilizing Compacted Clay against Chemical Attack," *Journal of Geotechnical Engineering,* October 1990.

12. Brown, R. W., "Concrete Foundation Failures," *Concrete Construction,* March 1968.

13. Brown, R.W., *Design and Repair Residential and Light Commercial Foundations,* McGraw-Hill, New York, 1990.

14. Brown, R. W., "A Field Evaluation of Current Foundation Design vs. Failure," *Texas Contractor,* July 6, 1976.

15. Brown, R.W., *Foundation Behavior and Repair,* 2d ed., McGraw-Hill, New York, 1992.

16. Brown, R.W., *Foundation Behavior and Repair,* 3d ed., McGraw-Hill, New York, 1997.

17. Brown, R.W. (ed.), *Practical Foundation Engineering Handbook,* McGraw-Hill, New York, 1995.

18. Brown, R. W., *Residential Foundations: Design, Behavior and Repair,* 2d ed., Van Nostrand Reinhold, New York, 1984.

19. Brown, R. W., "A Series on Stabilization of Soils by Pressure Grouting," *Texas Contractor,* Part 1A, January 19, 1965; Part 1B, February 2, 1965; Part 2B, March 16, 1965.

20. Brown, R. W., and Gilbert, B., "Pressure Drop Perforations Can Be Computed," *Petroleum Engineer,* September 1957.

21. Bruce, D. A., and Nicholson, P. J. "Minipile Mature in America," *Civil Engineer,* ASCE, December 1988.

22. Budge, W. D., et al., *A Review of Literature on Swelling Clays,* Colorado Dept. of Highways, Denver, 1964.

23. Building Research Establishment, "Minipiling for Low Rise Buildings," *Digest 313,* Garston, Waterford, England, September 1986.

24. Canadian Geotechnical Society, *Canadian Foundation Engineering Manual,* 2d ed., 1985.

25. Carter, M., and Bently, S. P., *Correlations of Soil Properties,* Pentech Press, London, 1991.

26. Chen, F. H., *Foundation on Expansive Soils,* Elsevier, New York, 1988.

27. Chen, F. H., "A Practical Approach on Heave Prediction," Seventh International Conference on Expansive Soils, ASCE, Dallas, 1992.

28. Choppin, N. J., and Richardson, I. G., *Use of Vegetation in Civil Engineering,* Buttersworth, London, 1990.

29. Conduto, Don P., *Foundation Design,* Prentice Hall, Englewood Cliffs, New Jersey, 1994.

30. Cozart, C. D., and Burke, Jack W., "Collapsible Soils in Texas," ASCE Section Meeting, Austin, Texas, April 1990.

31. Crilly, M. S., et al., "Seasonal Ground and Water Movement Observations from an Expansive Clay Site in the U.K.," Seventh International Conference on Expansive Soils, Dallas, 1992.

32. Davis, R. C., and Tucker, Richard, *Soil Moisture and Temperature Variation Beneath a Slab Barrier on Expansive Clay,* Report No. TR-3-73, Construction Research Center, University of Texas at Arlington, May 1973.

33. Day, R. W., "Sample Disturbance of Collapsible Soil," *Journal of Geotechnical Engineering,* ASCE, Vol. 116, January 1990.

34. Einstein, H. H., "Observations, Quantification and Judgement: Terzaghi and Engineering Geology," *Journal of Geotechnical Engineering,* November 1991.

35. Evans, J. C., and Pancoski, S. E., *Organically Modified Clays,* Paper No. 880587, Transportation Research Board, Washington, D.C., January 1989.

36. Federal Housing Administration, *Criteria for Selection and Design of Residential Slab-On-Ground,* Building Research Advisory Board, Report No. 33, Publication No. 1571, National Academy of Sciences, 1968.

37. Ferguson, Phil M., *Reinforced Concrete Fundamentals,* 3d ed., John Wiley, New York, 1973.

38. Foreman, D. E., et al., "Permeation of Compacted Clay with Organic Chemicals," *Journal of Geotechnical Engineering,* July 1986.

39. Fourie, A., "Laboratory Evaluation of Lateral Swelling Pressures," *JGE,* ASCE, October 1989.

40. Fredlund, D. C., et al., "Variability of an Expansive Clay Deposit," *Fourth International Conference on Expansive Soils,* ASCE, Denver, Colorado, 1980.

41. Freeman, T. J., et al., *Has Your House Got Cracks?,* Institute of Civil Engineers and Building Research Establishment, London, 1994.

42. Freeman, T. J., et al., "Seasonal Foundation Movements in London Clay," Fourth International Conference, University of Wales College of Cardiff, July 1991.

43. Gieseking, J. E., "Mechanics of Cation Exchange in the Montmorillonite-Beidilite-Nontronite Type of Clay Mineral," *Soil Science,* Vol. 4, 1939, pp. 1–14.

44. Greenfield, S. J., and Shen, C. K., *Foundation in Expansive Soils,* Prentice Hall, New Jersey, 1992.

45. Grim, R. E., *Applied Clay Mineralogy,* McGraw-Hill, New York, 1962.

46. Grim, R. E., *Clay Mineralogy,* McGraw-Hill, New York, 1953 and 1968.

47. Grim, R. E., and Guven, N., *Bentonites-Geology, Mineralogy, Properties and Uses,* Elsevier, New York, 1978.

48. Grubbe, Berry, "Lime Induces Heave of Eagle Ford Clays in Southeast Dallas County," ASCE Fall Meeting, El Paso, 1990.

49. Hall, Gerald, "Garden Questions—How to Get a Fruitful Apple Tree," *Dallas Times Herald,* March 24, 1989.

50. Haller, John, *Tree Care,* MacMillan, New York, 1986.

51. Handy, R. L., *The Day the House Fell,* ASCE Press, New York, 1995.

52. Holland, J. E., and Lawrence, C., et al., "The Behavior and Design of Housing Slabs on Expansive Soils," Fourth International Conference on Expansive Soils, ASCE, Denver, Colorado, 1980.

53. Holland, J. E., and Lawrence, C. E., "Seasonal Heave of Australian Clay Soils," Fourth International Conference on Expansive Soils, ASCE, Denver, Colorado, 1980.

54. House Building Council, *Building Near Trees,* Practice Note No. 3, London, 1985.

55. Houston, S. L., Houston, W. N., and Spadola, D. J., "Prediction of Field Collapse of Soils Due to Wetting," *Journal of Geotechnical Engineering,* ASCE, Vol. 114, No. 1, January 1988.

56. Houston, S. L., and Mostafa, El Ehwany, "Settlement and Moisture Movement in Collapsible Soils," *Journal of Geotechnical Engineering,* Vol. 116, October 1990.

57. Jones, D. E., and Holz, W. G., "Expansive Soils—The Hidden Disaster," *Civil Engineer,* ASCE, August 1973.

58. Jones, Earl D., and Jones, Karen A., "Treating Expansive Soils," *Civil Engineering,* August 1987.

59. Jordan, John W., "Alteration of the Properties of Bentonite by Reaction with Amines," *Journal of Mineralogical Society,* London, June 1949.

60. Kaminetsky, Dov, *Design and Construction Failures: Lessons from Forensic Investigations,* McGraw-Hill, New York, 1991.

61. Kaminetzky, Dov, "Rehabilitation and Renovation of Concrete Buildings," *Proceedings of Workshop,* National Science Foundation and ASCE, New York, February 1985.

62. Komornik, D., et al., *Effects of Swelling Clays on Piles,* Israel Institute of Technology, Haifa Israel.

63. Koslowski, T. T., *Water Deficits and Plant Growth,* Vol. 1, Academic Press, New York, 1968.

64. Krohn, J. P., and Slosson, J. E., "Assessment of Expansive Soils in the United States," Fourth International Conference on Expansive Soils, ASCE, Denver, Colorado, 1980.

65. Lambe, T. W., and Whitman, R. V., *Soil Mechanics,* John Wiley, New York, 1969.

66. Langenback, G. F., Apparatus for a Method of Shoring a Foundation, Patent No. 3,902,326, September 2, 1975.

67. Lawton, E. C., Fragaszy, R. J., and Hardcastle, J. H., "Collapse of Compacted, Clayey Sand," *Journal of Geotechnical Engineering,* ASCE, Vol. 115, September 1989.

68. McKeen, R. G., and Johnson, L. D., "Climate Controlled Soil Design—Perimeters for Mat Foundations," *JGE,* July 1990.

69. McKeen, R. G., "A Model for Predicting Expansive Soil Behavior," Seventh International Conference on Expansive Soils, ASCE, Dallas, 1992.

70. McLaughlin, H. C., et al., "Aqueous Polymers for Treating Clays," SPE Paper No. 6008, Society of Petroleum Engineers, Bakersfield, California, 1976.

71. McWhorter, David B., and Sunada, Daniel K., *Ground-Water Hydrology and Hydraulics,* Water Resources, Fort Collins, Colorado, 1977.

72. Meinzer, O. E., et al., *Hydrology,* McGraw-Hill, New York, 1942.

73. Meyer, Kirby T., "Defining Foundation Failure," Texas Section ASCE, Padre Island, Fall 1991.

74. Meyerhoff, G. G., "The Ultimate Bearing Capacity of Foundations," *Canadian Geotechnique Journal ,* Vol. 2, No. 4, 1951.

75. National House—Building Council, *Building Near Trees,* Practice Note No. 3, London, 1985.

76. Rep. NAVFAC DM-7.1 Naval Facilities Engineering Command, *Soil Mechanics, Design Manual 7.1,* Annapolis, Maryland.

77. Neely, W. J., "Bearing Capacity of Auger-Cast Pilings in Sand," *JGE,* ASCE, February 1991.

78. Nelson, John D., et al., "Moisture Content and Heave Beneath Slabs on Grade" and "Prediction of Floor Heave," Fifth International Conference on Expansive Soils, ASCE, Adelaide, South Australia, May 1984.

79. Nelson, J. D., and Miller, D. J., *Expansive Soils,* John Wiley Interscience, New York, 1992.

80. O'Neil, M., and Mofor, D., *Results of Concentric and Eccentric Loading,* University of Houston, Texas, 1988.

81. Orcut, R. G., et al., The Movement of Radioactive Strontium through Naturally Porous Media, Atomic Energy Commission Report, November 1, 1955.

82. Perrin, L. L., "Expansion of Lime Treated Clays Containing Sulfates," Seventh International Conference on Expansive Soils, ASCE, Dallas, 1992.

83. Petry, T. H., and Armstrong, J. C., "Geotechnical Engineering Considerations for Design of Slabs on Active Clay Soils," ACI Convention, Dallas, February 1981.

84. Petry, T. M., and Brown, R. W., "Laboratory and Field Experiences Using Soil Sta Chemical Soil Stabilizer," ASCE Paper, San Antonio, October 3, 1986; *Texas Civil Engineer,* February 1987, pp. 15–19.

85. Petry, T. M., and Little, D. N., "Update of Sulfate Induced Heave in Lime and Portland Cement Treated Clays," *1992* Annual Meeting of Transportation Research Board, Washington, D.C., August 1991.

86. Petry, T., Armstrong, C., and Poor, A., Foundation Seminar, University of Texas at Arlington, May 1983, and December 1985.

87. Pettit, P., and Wooden, C. E., "Jet Grouting: The Pace Quickens," *Civil Engineering,* 1988.

88. Pirson, S. J., *Soil Reservoir Engineering,* McGraw-Hill, New York, 1958.

89. Poor, Arthur, *Experimental Residential Foundations on Expansive Soils,* HUD Contract H-2240R, 1975–79, University of Texas at Arlington.

90. Post Tension Institute, *Design and Construction of Post-Tension Slabs-On-Grade,* 1st ed., Phoenix, Arizona, 1980, 1996.

91. Prakash, S., and Sharma, H. D., *Pile Foundations in Engineering Practice,* John Wiley, New York, 1990.

92. Raj, P. P., *Geotechnical Engineering,* Tata McGraw-Hill, New Delhi, India, 1995.

93. Ramsey, Dan, *Foundation and Floor Framing,* McGraw-Hill, 1995.

94. Readers Digest, *The Great American West,* 3d ed., 1987.

95. Seeyle, E. E., "Data Book for Civil Engineers," *Design,* Vol. 1, 2d ed., John Wiley, New York, 1945.

96. Sherif, M. A., et al., "Swelling of Wyoming Montmorillonite and Sand Mixtures," *Journal of Geotechnical Engineering,* January 1982, pp. 33–45.

97. Sowa, V. A., "Influences of Construction Conditions on Heave of Slab-on-Grade Floors Constructed on Swelling Clays," 38th Canadian Geotechnical Conference, September 1985.

98. Sperry, Neil, *Complete Guide to Texas Gardening,* Taylor Publishing, Dallas, 1982.

99. Terzaghi, K., and Peck, R. B., *Soil Mechanics in Engineering Practice,* John Wiley, New York, 1948.

100. Tomlinson, M. J., *Foundation Design and Construction,* 4th & 5th eds., Wiley Science, New York, 1969, 1986.

101. Tschebotarioff, G. P., *Foundation, Retaining and Earth Structure,* 2d ed., McGraw-Hill, New York, 1973.

102. Tucker, R. L., and Poor, A., "Field Study of Moisture Effects on Slab Movement," *Journal of Geotechnical Engineering,* ASCE, April 1978.

103. Tucker, R., and Davis, R. L., *Soil Moisture and Temperature Variations Beneath a Slab Barrier on Expansive Soils,* Report No. TR-3-73, Construction Research Center, University of Texas at Arlington, 1973.

104. Williams, L. H., and Undercover, D. R., "New Polymer Offers Effective, Permanent Clay Stabilization Treatment," Society of Petroleum Engineers, Paper No. 8797, January 1980.

105. Winterkorn, H. F., and Fang, H. Y., *Foundation Engineering Handbook,* Van Nostrand Reinhold, New York, 1975, 2d ed., 1991.

106. Wrench, B. P., and Geldenhuis, J. J., "Heave and Settlement of Soils Due to Acid Attach," Seventh International Conference on Expansive Soils, ASCE, Dallas, 1992.

107. Wu, T. H., et al., "Study of Soil-Root Interaction," *Journal of Geotechnical Engineering,* December 1988.

108. Zebovitz, S., Krizek, R. J., and Amatzidis, D. K., "Injection of Fine Sands with Very Fine Cements," *Journal of Geotechnical Engineering,* Vol. 115, No. 12, December 1989.

GLOSSARY

Certain definitions concerning pile technology were taken from *Design and Installation of Driven Pile Foundations,* H. W. Hunt.

Active zone   The depth of seasonal soil moisture variation. Sometimes referred to as the zone of seasonal fluctuations.

Activity   Ratio of plasticity index (PI) to percentage, by weight, of clay particles passing through 2-μm screen.

Adequate watering   Watering sufficient to stop or arrest settlement brought about by soil shrinkage resulting from loss of moisture.

Aeolian soil   Soil that has been deposited by wind.

Aeration zone   The capillary fringe, an intermediate belt (which may include one or more perched water zones), and at the surface, the soil water belt, often referred to as the root zone.

Allowable load   The load that may be safely transmitted to a foundation member or a bearing soil.

Allowable pier or pile load   The load permitted on any vertical pier or pile applied concentrically in the direction of its axis. It is the least value determined from the capacity of the pile or pier as a structural member, the allowable bearing pressure on soil strata underlying the tips, the resistance to penetration, the capacity demonstrated by load test, or the basic maximum load prescribed by the building code. The latter may be exceeded where a higher value can be substantiated on the basis of test analysis.

Alluvial soil   Soil deposition by running waters.

Amphipathic   Molecules that have both polar and nonpolar ends.

Anchor pier or pile   A pile or pier connected to a structure by one or more ties to furnish lateral support or to resist uplift. Also, a reaction pile or pier for load testing.

Arenaceous soil   Sandy soil.

Argillaceous soil   Clayey soil.

Augered or drilled pier or pile   A concrete pier or pile cast in place in an augered hole, which may be belled at the bottom. Particularly suitable where soil is dry and hole will stand open. Otherwise, casing is required.

Basal spacing   The distance between individual or molecular layers of the clay particles.

Batter pier or pile   A pile driven or a pier cast at an inclination from the vertical. Also referred to as raked.

Bearing capacity   The allowable pier or pile load as limited by the provision that the pressures in materials along the pier or pile and below the tips (produced by the loads on individual pier or piles and by the aggregate of all piers or piles in the group) shall not exceed the allowable soil-bearing values.

Bearing capacity of soil   The maximum pressure that can be applied to a soil mass without causing shear failure. The pressure or stress is created by applied loads and transmitted to the soil by the foundation.

Bell (underream)   Enlargement of bottom end of augered pier to increase bearing-load capacity, usually limited to a maximum of 3.0 times the shaft diameter.

Bentonite (montmorillonite)   A colloidal clay used as a heavy slurry to prevent earth sloughing into a drilled hole, for waterproofing, or to aid in removal of cuttings during drilling. A principal constituent in expansive soils.

Brace pier or pile   A batter pile or pier connected to a structure in a way to resist lateral forces.

Caliche   Argillaceous limestone or calcareous clay.

Canopy (width)   The diameter of tree foliage; can be referred to as drip line.

Capillary fringe   An area that contains capillary water originating from the water table or perched water zone.

Capillary pressure   With reference to soils, the negative water pressure at points above the water table. Sometimes defined as the difference between air and water pressures in the pore space.

Capillary rise (h_c)   A measure of the height of water rise above the level of the free water boundary. Capillarity is impeded by the swell of clay particles (loss of permeability) upon invasion of water. Finer soils will create a greater height of capillary rise, but the rate of rise is slower. The result of surface tension.

Cast-in-place pier or pile   A concrete pier or pile poured (with or without a metal casing) in its permanent location in the ground.

Catema   Soil series originating from the same parent material and differing only in soil profile as a result of topographical conditions (erosion, percolation).

Clay   A soil that has the finest possible particles, usually smaller than $\frac{1}{10,000}$ in in diameter (2.5×10^{-4} cm) and often possessing the capacity for extreme volume changes with differential access to water.

Clay bearing failure   The result of expansive soils exerting nonuniform pressure against a constant downward loading. Such a loading causes the pier to deviate further away from vertical until the pier can no longer support a structural load. The latter is also referred to as "failure in lateral stress."

Cohesion   A cementing or gluing force between particles; requires a clay content.

Collapsible soil   Soil susceptible to substantial reduction in void ratio upon addition of water.

Compacted concrete pier or pile   A cast-in-place pier or pile formed with an enlarged base. Concrete in the base is placed in small batches, which are compacted by heavy blows prior to attaining an initial set.

Composite pile   A pile made up of two types of piles joined together. The connection between the two components should prevent their separation during and after construction. The connection is sometimes referred to as a bond beam.

Compressive stress   Stress resulting from compressive loading, considered positive in soil mechanics.

Consolidation   The action that occurs when applied load forces water from voids, enabling solid particles to become more closely packed. The resultant downward movement is termed *settlement.*

Corbelling   The practice in Victorian times of spreading a foundation load over a greater area by stepping the brick footing.

Covalent bond   Chemical bond resulting from the sharing of a pair of electrons, one supplied by each atom forming the bond.

Cushion block   Material inserted between the hammer and the pile-driving cap or pile to minimize local damage. Asbestos, micarta, steel, aluminum, coiled cable, wood, and other materials are used.

Cut and fill   Removal of excess existing soil (cut) to low or deficient areas (fill) for contouring purposes.

Cutting shoe   Additional metal, such as an inside or outside cast steel ring or welded plate at the bottom of an open-end pile or caisson, to strengthen the tip.

Cutoff   The prescribed elevation at which the top of a driven pile is cut or left; also, the portion removed from the upper end of the pile after driving.

Dead load   The weight of the empty structure.

Deep foundation   A design whereby structural load is transmitted to a soil at some depth, usually through piers, piles, or caissons.

Dipole   Two equal and opposite charges separated in space.

Displacement pile   A solid timber or precast concrete pile or hollow pile driven with the lower end closed, which displaces soil volume by compaction or by lateral or vertical displacement of soil. H and open-end pipe are examples of nondisplacement piles.

Downdrag   Negative friction or weight of earth gripping a pile in settling soils and thus adding load on the piles installed.

Drift   Materials deposited by glacial movement.

Drilled-in caisson   An open-ended pipe driven to rock. It is cleaned out and a socket drilled into rock to receive a steel core (H, WF, or bars). Finally, the socket and pipe are filled with concrete.

Driving cap or helmet   A steel cap placed over a pile to prevent damage in driving.

Earth anchor   A steel shaft containing one or more helixes that is screwed into the earth to provide a retention system against uplift forces.

Effective size (D_{10})   Sand particle size such that 10 percent passes that sieve.

Elastic modulus (E)   A factor determined by dividing stress by strain, expressed in pounds per square inch, or kilograms per square centimeters.

Elevations   Measurements taken by instrument (usually optical) to establish grades.

Elutriation   The act of purification by washing and straining or decanting.

Embedment   The length of the pile from the surface of the ground, or from the cut-off below the ground, to the tip of the pile. Also, the depth of penetration of the top of the pile into the pile cap.

Ettringite ($3\,CaO \cdot Al_2O_3 \cdot 3CaSo_4 \cdot 32\,H_2O$)   The by-product of lime-sulfate-clay reaction. Subject to produce uncontrolled heave of soils involved.

Failure of pier or pile foundation   The movement of the pier or pile foundation, or any part thereof, either as settlement upheaval or laterally, to such an extent that objectionable damage results to the structure supported by the foundation. Also, failure of a pier or pile (or piers or piles) to pass a load test.

Field capacity   The residual amount of water held in the soil after excess gravitational water has drained and after the overall rate of downward water movement has decreased (zero capillarity).

Fill   Soil added to provide the level or desired construction surface or grade.

Fineness modulus   Sum of percentages of aggregates larger than each of the standard sieve sizes from No. 100 on, expressed as a decimal.

Flexural strength   The maximum stress (tensile or compressive) at rupture. (A fictitious, although sometimes convenient, value because mathematical assumptions for soil behavior at stresses approaching failure are inadequate.) Also referred to as modulus of rupture.

Footing   A member, usually concrete, that distributes the foundation load over an extended area and thus provides increased support capacity on any bearing soil.

Foot, pile   The lower end of a driven pile.

Force   Pressure times area or mass times acceleration.

Foundation   The part of a structure in direct contact with the ground that transmits the load of the structure to the ground.

Franki pile   A pile formed by ramming very dry concrete into the ground with a heavy weight, a pressure-injected footing.

Free water   Water that can be taken on or lost by the soil without corresponding soil volume change.

French drain   A perforated pipe installed in a cut to intercept and divert the underground water. The cut is below the level of the intruding water and is graded to drain the accumulated water away from the site. Sometimes a catch basin and discharge pump are required if sufficient natural grade does not exist.

Friction pile   A pile or pier that supports its load by the friction developed between the surface of the pile or pier and the soil through which it is driven.

Frost heaving   Expansion that results when a mixture of soil and water freezes. Upon freezing, the total volume may increase by as much as 25 percent, dependent upon the formation of ice lenses at the boundary between the frozen and unfrozen soil.

g_T Unit weight of liquid at temperature T, expressed in grams per cubic centimeters.

Gap-graded soil   A coarse-grained soil containing both large and small sizes but a relatively low proportion of intermediate sizes.

Geopile   Short, densely compacted aggregate (usually highway base coarse material) pier designed to enhance a soil through bottom densification with horizontal and vertical prestressing.

Grade   The level of ground surface. Also, the rise or fall per given distance (often 100 ft, or 30 m).

Grout curtain   A continuous, consolidated boundary or area with strength sufficient to permit excavation without sloughing or to provide adequate bearing strength from soils of basic substandard capacity.

Grouting   An operation whereby a material is injected to penetrate and permeate a relatively deep soil bed. The purpose is to decrease voids or permeability, increase strengths of the penetrated soil, or, on occasion, impede organic decay.

Guard pile   A fender pile.

Guide pile   A pile used as a guide for driving other piles.

Guides   The part of the pile leads that forms a pathway for the hammer. It consists of parallel members that mate with grooves on the hammer.

Gumbo   Highly plastic clay from the southern and/or western United States.

Gypsum   Hydrous calcium sulfate ($Ca\ SO_4$).

Hammer energy   The capacity of a pile-driving hammer to do work at impact, measured in foot·pounds per blow.

Heave of pier or pile   The uplift of earth between or near piers or piles caused by the displacement of soil by pile driving; the uplift of an in-place pile caused by the driving of an adjacent pile; the uplift of an in-place pile or pier caused by cohesive skin friction (expansive soils).

Hydraulic radius   Cross-sectional area of flow divided by wetted perimeter.

Hydrogenesis   Moisture condensed from air by cooling soil temperature.

Hydrophobic   Nonpolar end of molecule is water insoluble (water-fearing).

Hydrophylic   Polar end of molecule is water soluble (water-loving).

Hygroscopic soil   Soil that readily takes up and retains water on the surface of soil particles.

Intercalation   Absorption due to pillaring of clay by surfactant.

Interception   The process whereby precipitation is caught and held by foliage and evaporated from the exposed surfaces.

Interior floors   Floors that are supported by the girder-and-joist system of a wood substructure wherein the system is in turn supported on piers and pier caps.

Interlayer moisture   Water that is situated within the crystalline layers of the clay and provides the bulk of the residual moisture contained within the intermediate belt.

Intermediate belt   Soil that contains moisture essentially in dead storage and is held by molecular forces. It may include one or more perched water zones.

Interstitial water   *See* Pore water and Interlayer moisture.

Ionic bond   Chemical bond resulting from the complete transfer of electrons from one atom to the other.

Jacking   A means of imposing a static driving force on a pile by jacks. Used extensively to install piles in underpinning existing structures and in static load testing.

Jetting   The use of a water jet to facilitate the placing or driving of a pile through the hydraulic displacement of parts of the soil. In some cases, a high-pressure air jet may be used either alone or with water. Sometimes used to facilitate consolidation.

Lacustrine   Soils deposited in glacial lakes.

Lateral support   Batter piles or reinforcement to resist lateral forces on piles or footings.

Laterites (brick)   Soils developed from intense leaching in well-drained soils of silicates and bases under tropical and subtropical conditions. The soil can be cut with iron tools but upon exposure to air becomes "brick" hard.

Liquid limit   The water content in a soil at which 25 blows from a drop of 1 cm closes a $^3/_8$-in V groove for a length of $^1/_2$ in in a standard liquid limit (LL) device.

Live load (W_L)   The weight of building contents plus wind, snow, and earthquake forces where applicable.

Loess   An aeolian deposit of uniform gradation with some calcareous cementation.

Mandrel   A core that is inserted into a closed-end thin-shell tubular pile. A solid mandrel is heavy tubular section that will transmit the hammer energy to the point; a collapsible mandrel is a core, the outer diameter of which can be changed by mechanical or other means, that is capable of transmitting the hammer energy to the bottom of the pile and supporting the wall of thin-shelled casing. It is inserted into the pile in a collapsed condition and expanded to grip the inner surface of the pile with sufficient force to prevent slipping. After the pile is in place, the mandrel is collapsed and withdrawn.

Marl   A calcareous clay.

Maximum density   The density attained by the addition of sufficient water to fill the voids and help the particles move closer together. Water added beyond that point displaces the heavier solids and thus reduces the density.

Minipile   A pile with a very small diameter-to-length ratio that transfers loads almost entirely by skin friction construction; may be steel reinforced concrete (grout), steel, or sometimes wood.

Moisture barrier   A means of maintaining moisture content beneath a foundation (generally slab) consisting of an impermeable barrier extending to some depth and in close proximity to the perimeter beam.

Mudjacking   A process whereby a water and soil cement or soil-lime-cement grout is pumped beneath the slab, under pressure, to produce a lifting force that literally floats the slab to desired position.

Negative friction   The effect on a pile of settling soil that may grip the pile and add its weight to the load to be carried by the bearing strata.

Newton   A unit of force: $1 \text{ N} = 1 \text{ kg (mass)} \times 1 \text{ m/s}^2$ (acceleration).

Newtonian fluid   A fluid (e.g., water, antifreeze, salt water) that produces a straight line through the origin when a shear rate–shear shear stress diagram is plotted.

Noncohesive soil   A soil in which there is no attraction or adhesion between individual soil particles.

Nonnewtonian fluid   A fluid (e.g., cement, soil, or soil cement slurry) that does not exhibit a linear shear rate–shear stress diagram.

Organophilic   A substance in which the polar end is soluble in organic materials (organic-loving, water-fearing).

Organophobic   A substance in which the nonpolar end of molecule is insoluble in organic materials (organic-fearing).

Osmosis   The transfer of water through a semipermeable membrane. The increased pressure caused by the diffusion of water is referred to as the osmotic pressure.

Peak flow   Maximum instantaneous rate of flow that occurs during a day. Sewer systems are specifically designed to handle this peak flow.

Perched water zones   The region of perched ground water that develops essentially from water accumulation either above a relatively impermeable stratum or within an unusually permeable lens. It generally occurs after a good rain and is relatively temporary.

Permafrost   Refers to that condition where the subsoil remains continuously frozen—involves perennially frigid areas.

Perma-Jack   A proprietary device (1976) that utilizes a hydraulic ram to drive slip-jointed sections of 3-in steel pipe into rock or suitable bearing. Suffers a number of recent "knock-offs."

pH   A measure of hydrogen ion concentration by which 7 is neutral. Values less than 7 indicate acidity, and values above 7 indicate basicity.

Phreatic boundary   The surface of the water table that will not normally deflect or deform (excepting infrequent conditions) in the proximity of a producing well.

Pier and beam   A design wherein the perimeter loads are carried on a continuous beam supported in turn on piers drilled into the ground, supposedly to a competent bearing soil or stratum. Interior loads are carried by isolated piers in a grid pattern.

Pier and beam, low profile   A design wherein the crawl space is substantially lower than the exterior grade.

Pier and beam, normal design   A design wherein the crawl space is at a grade equivalent to the exterior landscape.

Piers   In general, concrete poured into circular holes that (as opposed to piles) are normally of larger cross-sectional area relative to length. Shafts are generally extended through marginal soils to either rock or competent bearing material. For residential and lightly loaded structures the optimum pier diameter has been established to be 10 to 12 in.

Pile   Long, slender wood, steel, or concrete members usually driven in groups or clusters. They may also be poured concrete, which gives rise to the gray area of differentiation between piles and slim piers.

Pile bent   Two or more piles driven in a row transverse to the long dimension of a structure, such as a bridge foundation, and fastened together by capping and bracing.

Pile bulkhead   A pile structure generally consisting of vertical piles, with brace or anchor piles, wales, and a sheet pile wall, framed together and capable of resisting earth or water pressure.

Pile butt   The larger end of a tapered pile; section of pile cut off at top.

Pile-driving cap   Generally a forged steel or a steel casting designed to fit over and around the top or butt end of the pile to prevent damage to the head of the pile while driving it. Also known as a hood or bonnet.

Pile extractor   A device for pulling piles out of the ground. It may be an inverted steam or air hammer with yoke so equipped as to transmit upward blows to the pile body or a specially built extractor utilizing the same principle. Vibratory hammers may be especially effective. All extractor operations require a strong upward force.

Pile hammer   One of several devices: (1) Drop hammer: A heavy weight, usually a metal casting with grooves in the sides to mate with leads. It is raised in the leads by ropes or cable and allowed to drop on the pile. Sometimes called a gravity hammer. (2) Steam or air hammer: A movable ram attached to a piston, operating in a cylinder, which in turn is mounted in a metal frame with grooves that engage the pile driver leads; it has a hood or bonnet on the lower end with cushion block that fits the pile head. In the single-acting hammer, steam or air pressure is used only to raise the moving parts, which then fall by gravity and strike the cushion block. In the double-acting and differential-acting hammer, steam or air pressure is used also to accelerate the downward movement of the ram. (3) Diesel hammer: An integrally powered pile hammer operated with the use of diesel fuel oil. A movable ram is raised initially by outside means. When released, the ram falls onto an impact block or anvil that itself rests on the pile. The falling ram actuates ports for admission of fuel and air, compresses the air, and raises the air to a higher temperature; the oil vaporizes and ignites from heat and pressure. The instantaneous expansion of gases gives additional raises the ram for the next stroke. (4) Vibratory driving machine: A unit with eccentric weights mounted on shafts rotated in different directions to apply periodic unbalanced forces to a pile shaft. An electric motor or hydraulic fluid rotates the shaft at about 700 to 2000 vibrations per minute. (5) Sonic driver: A machine designed for variation from 3600 to 9000 cycles per minute to match the sonic response of soil for very rapid driving.(Vibratory and sonic drivers work very well under some conditions but not efficiently in others. They generally produce little disturbance in surrounding soils.)

Pile or pier cap A structural member placed upon the tops of piers or piles. It is used to transmit and distribute the load of the structure above to the head of a pier, a pile, a row of piles, or a pile group.

Pile point A cast-steel or steel drive shoe, which may be pointed and is fixed to the pile shaft at the tip for easier driving and improved penetration. It also protects against damage from dense material or boulders and improves bearing at the tip.

Pile splicer A metal fitting for quickly joining two (similar or dissimilar) parts of a pile, in or out of the leads.

Pile tip The lower and usually the smaller end of a pile.

Pillaring Action that increases porosity by forcing clay platelets apart.

Pin pile A small-diameter drilled and grouted pile. *See also* Minipile.

Pipe flow apparatus The simplest capillary viscometer. The apparatus times the passing of a given volume of liquid.

Pipe pile A steel or concrete cylindrical shell of specified strength and thickness. It is driven either open or closed ended and is usually filled with concrete.

Plastic limit (PL) The water content at which a $1/8$-in-diameter thread of soil begins to crumble.

Plasticity index (PI) A dimensionless constant that bears a direct relationship to the affinity of the soil for volumetric changes with respect to moisture variations. The PI is determined as the difference between the liquid limit (LL) and plastic limit (PL).

Plates Wood members placed horizontally at each end of wall studs. The ceiling plate, or header, forms the top, and the floor or sole plate forms the bottom.

Polyurethane A by-product produced by reacting a dihydroxy alcohol with a diisocyonate.

Poorly graded soil A coarse-grained soil in which a majority of particles are of one size. Often described as uniform or gap-graded.

Pore water Water that occurs within the soil mass external to individual soil grains and is held by interfacial tension.

Porosity (*n*) The ratio of combined volume of liquid and air to total volume of soil.

Precast pier or pile A reinforced concrete pier or pile that is manufactured in a construction yard or at the site and, having been properly cured, is driven like a timber pile.

Preexcavation Removal of soil that may heave. The removal is effected by auguring or by driving and cleaning out an open-end pipe. The objective is to keep soil volume constant.

Prestressed pier or pile A precast concrete pier or pile that is either pretensioned or post-tensioned to reduce or eliminate tensile stresses to which piers or piles are subjected during transportation and driving and while in service. Pretensioning is used for piling up to about 24 in in size for which the wires are stressed prior to casting concrete around them. Post-tensioning is used for large open cylindrical piles; wires are pulled through holes left in the walls and stressed against concrete previously cast in the walls.

q_u The unconfined compressive strength that measures a soil's capacity to carry load.

r The radius of a capillary pore, conduit, pier, pile, etc.

Rake angle   Angle off vertical; *see* Batter pier or pile.

Refusal   The condition reached when a pile being driven by a hammer has zero penetration per blow (as when the point of the pile reaches an impenetrable bottom such as rock) or when the effective energy of the hammer blow is no longer sufficient to cause penetration. When so stipulated, the term *refusal,* or *substantial refusal,* may be used to indicate the specified minimum penetration per blow. Overdriving after essential refusal can result in serious damage to the pile.

Regolith   The region of uncemented mineral matter above the bedrock.

Rock   In the construction field, a layer of stoney material. Contrary to popular opinion, it is not always a superior foundation bed, depending upon such possible factors as the presence of bedding planes, faults, joints, weathering, and cementation (or lack thereof) of constituents.

Root zone   The upper layer of soil from which plant roots take their moisture.

Run-off   Excess water not retained by the soil.

Sand pile   A column of sand installed by driving a pipe with an openable end into soil (usually swamp) and then forcing the sand out as a permanent column as pipe is withdrawn. The sand serves as a drain or wick to speed consolidation and improve bearing value of the soil.

Sands and gravels   Coarse particles that range in size from 3 in (7.5 cm) in diameter down to grains so small they can be barely distinguished by the unaided eye.

Sanitary landfill   A euphemism for garbage dump.

Saprolites   Soils developed from in situ weathering of rocks.

Saturation zone   The deepest soil water source. More commonly termed the *water table* or *groundwater.*

Screw pile   A spiral blade fixed on a shaft and screwed into the ground by a rotating force. Quite large piles of this type have been used, but they are little known in the United States.

Sensitivity   The ratio of unconfined compressive strength of undisturbed clay to that of remolded clay.

Set   Net penetration per pile-driving blow.

Settlement   The drop of some portion of the foundation below the original as-built grade.

Settling velocity (V_s)   The rate at which particles settle in a liquid, described by Stoke's law.

shale   A sedimentary rock; indurated clay and/or silt muds.

Shallow foundation   A foundation in which the depth (D) is less than or equal to the width (B). Refers principally to footings.

Shear strength of soil (coulomb) $\tau_f = C + \alpha \tan \phi$ (cohesion plus friction), lb/in^2 (kg/cm^2), where S or τ = shearing resist once C = apparent cohesion, α = total normal stress on shear plane, and ϕ = angle of shearing stress. (Sometimes S is used rather than τ.)

Shearing strength of clay soil (vane shear test):

$$C = \frac{T}{[(d^2 h/2) + (d^3/6)]} \qquad \text{lb/in2/(kg/cm2)}$$

where, d = vane width, h = vane height, and T = torque required to rotate vane1.

Sheet pile A pile that may form one of a continuous line or row of timber, precast concrete, or steel piles driven in close contact to provide a tight wall. Used to resist or prevent lateral pressure caused by water, adjacent earth, or other materials.

Shrinkage limit (SL) Water content below which there is no volume decrease with further decreases in water content.

Silts Soil that is intermediate in particle size between sand and clay. It represents particles of ground rocks that have not yet changed in character to minerals.

Slab One or another variety of concrete foundation generally supported entirely by surface soils. It probably constitutes the majority of new construction in areas with high-clay soils.

Sliding The consequence of erecting a structure on a slope such that movement is not limited to vertical but has a lateral or horizontal component.

Soil All the loose material constituting the earth's crust. Constituents exist in varying proportions including air, water, and solid particles. The solid particles have largely been formed by the disintegration of different rocks.

Soil belt The vertical section that can contain capillary water available from rains or watering. Unless this moisture is continually restored, the soil will eventually desiccate through the effects of gravity, transpiration, and/or evaporation. That soil section that generally provides moisture for the vegetable and plant kingdom.

Soil stabilization A procedure for improving the natural soil properties to make them a more adequate base for construction.

Soil suction *See* Suction pressure.

Solum The upper region in a vertical soil profile. The zone studied by pedologists.

Spreadfootings Foundation supports that generally consist of two structural components: (1) steel-reinforced pad of sufficient size to adequately distribute the foundation load over the supporting soil and poured to a depth relatively independent of seasonal soil moisture variation and (2) a steel-reinforced pier cap tied into the footing with steel and poured to the bottom of the foundation beam. One of the principal designs used to underpin distressed foundations.

Spud A short, strong member driven and then removed (1) to make a hole for inserting a pile that is too long for placing directly in the pile driver leads or (2) to break through a crust of hard material.

Steel pipe pile   Pipe with any wall thickness or diameter. It may be driven open or closed end depending on conditions. Closed-end piles preferably have cast steel points but sometimes have flat plates. Open-end piles may require bottom reinforcement.

Step-tapered pier or pile   A cast-in-place pier or pile that has a diameter stepped up in increments, usually in 8- to 12-ft sections of corrugated thin-metal shell, and is driven with a mandrel. It may be pipe reduced in steps and installed with or without a mandrel.

Strength   The ability to resist force, often measured by the force (stress) required to cause rupture or failure. The basic resistance to failure is due to both cohesion and internal friction. The angle of rupture is a function of the intended angle of friction v. The angle that the plane of failure makes with the axis of loading is considered as equal to $45° + \phi/2$. Rupture or failure can result from applied tensile stress (failure in cohesion), shearing stress (sliding), compression stress (crushing), or some combination of the three.

Stress   The force at a point in a soil mass that is due to the weight of the soil above the point plus any applied (structural) load.

Stringer   A member at right angles to and resting on girders or in rare cases on pile or pier caps. Forms a support for the superstructure. (Also referred to as joists.)

Strut   A compression member that extends horizontally from bent to bent, or pile to pile in a bent, and serves as a stiffening member.

Suction pressure   The combined attraction of capillary and osmotic pressures to transport water.

Surface absorbed water   Water that occurs within the soil mass externally to individual soil grains and is held by molecular attraction between the clay particle and the dipolar water molecule.

Swelling pressure   Pressure required to prevent swell as soil takes on moisture above a stable moisture level.

T_{ST} Surface tension of liquid at temperature T.

Tensile stress   Stress resulting from tensile forces, considered negative in soil mechanics.

Thornwaite index   The amount of water that would be returned to the atmosphere by both evaporation from the ground and surface and transpiration by plants if there was an unlimited supply of water to plants and soil. A positive value indicates a net surplus, whereas a negative number indicates a net soil moisture deficit. As an example, Dallas is considered to be -1; Houston, $+18$; El Paso, -40; Seattle, $+100$; and Bangor, $+80$. The Thornwaite index is related to soil suction and moisture velocity.

Till   Materials deposited by ice.

Transpiration   The removal of soil moisture by vegetation.

Tremie   A procedure and equipment for placing concrete under water or to some distance below entry.

Triaxial shear (compression) strength test   A procedure in which a cylindrical sample, generally with a length/diameter ratio of 2, is stressed under conditions of axial symmetry. The sample is subjected to peripheral hydraulic pressure and then the axial stress is applied in gradually increasing increments by applications of compressive load until failure of the sample occurs.

Triaxial tests   Tests that are basically useful for clays, silts, peats, and soft rocks: (1) Unconsolidated-undrained: Soil water is retained within the sample during the test. Consolidation of sample by peripheral fluid pressure is not allowed. (2) Consolidated-undrained: Drainage of soil water from specimen is permitted under a specified peripheral fluid pressure until consolidation is complete. Drainage is blocked and the axial stress applied. Pore water pressures can be measured during the undrained portion of test. (3) Drained: Drainage of specimen is permitted through duration of test. The principal stress (axial load) is applied at a rate sufficiently low to ensure that the water pore pressure remains zero. Under conditions in which the test procedures are reflective of field conditions, the undrained strength can be expressed in terms of total stress (C and ϕ) and the drained strength in terms of effective stress (C' and ϕ').

Trip   A block in the leads of a pile driver of the drop hammer type that has a device for releasing the hammer and thereby regulating the height of the fall; it is also used to drop the ram to start a diesel hammer.

Uniform soil   Soil that contains a high proportion of particles with narrow size limits.

Uniformity coefficient (C_u)   The ratio D_{60}/D_{10}, where each term represents grain size, respectively, that which passes 60 percent (D_{60}) and 10 percent (D_{10}) by weight.

Upheaval   The situation in which areas of the foundation (usually internal) raise above the as-built position.

Vadose   Pertains to water located in the aeration zone. *See also* Perched water zones.

Varve   A soil consisting of thin alternating layers of silt and clay. Deposited by water.

Viscosity   A single constant that describes the newtonian relationship of shear stress and shear rate.

Void ratio (e)   The ratio of combined volume of water and air to the total volume of the soil sample.

W_T The sum of live loads (W_L) and dead loads (W_D).

Water leaks   Water from any domestic source that accumulates under the foundation. Any water under the foundation, regardless of source, tends to accumulate in the plumbing ditch. Usually of greater concern to slab foundation.

Water table   The upper surface of water saturation in permeable soil or rock.

Well graded   A soil with a fairly even distribution of grain sizes—no excess of one size and no intermediate sizes lacking.

INDEX

ABOUT THE AUTHOR

Robert Wade Brown is an internationally respected expert on foundation problems and expansive soils, and the author of six prior books on these subjects. A physical chemist and petroleum engineer, he has developed systems that stimulate oil well production. He owns Brown Foundation Repair and Consulting, Inc., in Garland, Texas, and consults around the world through Brown Consolidated, Inc.